ICME-13 Monographs

Series editor

Gabriele Kaiser, Faculty of Education, Didactics of Mathematics, Universität Hamburg, Hamburg, Germany

Each volume in the series presents state-of-the art research on a particular topic in mathematics education and reflects the international debate as broadly as possible, while also incorporating insights into lesser-known areas of the discussion. Each volume is based on the discussions and presentations during the ICME-13 conference and includes the best papers from one of the ICME-13 Topical Study Groups, Discussion Groups or presentations from the thematic afternoon.

More information about this series at http://www.springer.com/series/15585

Werner Blum · Michèle Artigue ·
Maria Alessandra Mariotti ·
Rudolf Sträßer · Marja Van den Heuvel-Panhuizen
Editors

European Traditions
in Didactics of Mathematics

 Springer Open

Editors
Werner Blum
Department of Mathematics
University of Kassel
Kassel, Hesse, Germany

Maria Alessandra Mariotti
Department of Information Engineering
and Mathematics
University of Siena
Siena, Italy

Marja Van den Heuvel-Panhuizen
Freudenthal Institute/Freudenthal Goup
Utrecht University
Utrecht, Netherlands

and

Nord University
Bodø, Norway

Michèle Artigue
University Paris Diderot—Paris 7
Paris, France

Rudolf Sträßer
Institut für Didaktik der Mathematik
University of Giessen
Giessen, Hesse, Germany

ISSN 2520-8322 ISSN 2520-8330 (electronic)
ICME-13 Monographs
ISBN 978-3-030-05513-4 ISBN 978-3-030-05514-1 (eBook)
https://doi.org/10.1007/978-3-030-05514-1

Library of Congress Control Number: 2018964682

This Springer imprint is published by the registered company Springer Nature Switzerland AG
The registered company address is: Gewerbestrasse 11, 6330 Cham, Switzerland

Contents

Contents

Chapter 1
European Didactic Traditions in Mathematics: Introduction and Overview

Werner Blum, Michèle Artigue, Maria Alessandra Mariotti, Rudolf Sträßer and Marja Van den Heuvel-Panhuizen

Abstract European traditions in the didactics of mathematics share some common features such as a strong connection with mathematics and mathematicians, the key role of theory, the key role of design activities for learning and teaching environments, and a firm basis in empirical research. In this first chapter, these features are elaborated by referring to four cases: France, the Netherlands, Italy and Germany. In addition, this chapter gives an overview on the other chapters of the book.

Keywords European didactic traditions · Overview

1.1 Introduction

Across Europe, there have been a variety of traditions in mathematics education, both in the practice of learning and teaching at school and in research and development, which have resulted from different cultural, historical and political backgrounds. Despite these varying backgrounds, most of these traditions share some common

W. Blum (✉)
Department of Mathematics, University of Kassel, Kassel, Germany
e-mail: blum@mathematik.uni-kassel.de

M. Artigue
University Paris Diderot—Paris 7, Paris, France

M. A. Mariotti
Department of Information Engineering and Mathematics Science,
University of Siena, Siena, Italy

R. Sträßer
Institut für Didaktik der Mathematik, University of Giessen, Giessen, Germany

M. Van den Heuvel-Panhuizen
Utrecht University, Utrecht, The Netherlands

M. Van den Heuvel-Panhuizen
Nord University, Bodø, Norway

© The Author(s) 2019
W. Blum et al. (eds.), *European Traditions
in Didactics of Mathematics*, ICME-13 Monographs,
https://doi.org/10.1007/978-3-030-05514-1_1

1

features, one feature being the use in many languages of the word *didactic* (derived from the Greek *didáskein*, which means teaching) to denote the art and science of teaching and learning (*didactiek* in Dutch, *didactique* in French, *didáctica* in Spanish, *didattica* in Italian, *didaktika* in Czech, *dydaktyka* in Polish and *didaktik(k)* in Swedish, Danish, Norwegian and German) rather than *education*, which is preferred in Anglo-Saxon traditions. These European didactic traditions can be traced back as far as Comenius' *Didactica Magna* in the 17th century, the first comprehensive opus on aims, contents and methods of teaching. These traditions share in particular the following common features: a strong connection with mathematics and mathematicians, the key role of theory, the key role of design activities for learning and teaching environments, and a firm basis in empirical research. Other common features (such as the important role of proofs and proving or of linking mathematics with the real world) can be considered part of those four features.

In the following sections, we will elaborate a bit more on these four common features.[1] They will be made more concrete by referring briefly to four selected cases of European traditions in the didactics of mathematics: France, the Netherlands, Italy and Germany. In the following four chapters (Chaps. 2–5) of this volume, these four traditions are presented in considerable detail. In particular, the role of the four key features in those traditions will become more transparent. The last two chapters are devoted to another two European traditions, the Scandinavian (Denmark, Norway and Sweden; Chap. 6) and the Czech/Slovak (Chap. 7). Although this chapter is written in English, we will speak, throughout the chapter, of the 'didactics of mathematics' instead of 'mathematics education' when we refer to the discipline dealing with all aspects of teaching and learning mathematics in the above-mentioned European traditions.

1.2   The Role of Mathematics and Mathematicians

Here we will highlight the role that some outstanding mathematicians have played in the didactics of mathematics in these four countries by their involvement in educational issues such as designing curricula for school and for teacher education and writing textbooks and by their fostering of the development of didactics of mathematics as a research field. In this respect, a prominent exemplar is Felix Klein (see Tobies, 1981), who also had a great influence on other mathematicians who had the opportunity of getting to know his work during their visits to Germany as researchers.

An important occasion for international comparison of different experiences in the didactics of mathematics was the Fourth International Congress of Mathematicians, which took place in Rome from 6 to 11 April 1908. During this congress, the International Commission on the Teaching of Mathematics (Commission Internationale de l'Enseignement Mathématique, Internationale Mathematische Unterrichtskom-

---

[1] These sections refer to the corresponding sections in Blum, Artigue, Mariotti, Sträßer, and Van den Heuvel-Panhuizen (2017).

mission and Commissione Internazionale dell'Insegnamento Matematico in France, Germany and Italy, respectively) was founded (details of the history of this institution can be retrieved at http://www.icmihistory.unito.it/timeline.php).

After the traumatic interruptions for the First and Second World Wars, mathematicians were again involved in various reform movements. In many countries, the ideas and principles of the so-called New Math were shared in the 1960s and 1970s. We can recognise a common interest in reforming curricula, which is certainly related to the impact a new generation of mathematicians had on the reorganisation of mathematics that was initiated by the Bourbaki Group. Thus, although the concrete results of the New Math movement were very different in various countries, a common feature was that substantial innovation entered into school practice through the active involvement of eminent figures such as Gustave Choquet, Jean Dieudonné and André Lichnerowicz in France; Emma Castelnuovo in Italy; and Hans Freudenthal in the Netherlands.

In the context of this reform, new perspectives developed, beginning in the late 1970s, that moved the focus of reflection from issues concerning mathematical content and its organisation in an appropriate curriculum to issues concerning the description and explanation of the learning and teaching of mathematics, giving birth to a new scientific discipline, the didactics of mathematics, that rapidly developed through active international interaction. In some cases, for instance in France and Italy, it is possible to recognise again the strong influence of the mathematicians' community, since the first generation of researchers in the didactics of mathematics consisted in these countries for the most part of academics affiliated with mathematics departments. This observation does not ignore the existence of a recurrent tension between mathematicians and researchers in didactics of mathematics.

In summary, some common features that can be considered the core of the European tradition of didactics of mathematics can be directly related to the fruitful commitment of mathematicians to educational issues and their intent to improve the teaching and learning of mathematics. One example is the strong role that proofs and proving have in these European traditions. It can be said that in all four cases, mathematics has been and still is the most important related discipline for the didactics of mathematics, and there is still a lively dialogue between mathematicians and didacticians (researchers in the didactics of mathematics) on educational issues.

1.3 The Role of Theory

The word *theory* in the didactics of mathematics has a broad meaning, ranging from very local constructs to structured systems of concepts; some are 'home-grown' while others are 'borrowed' with some adaptation from other fields, and some have developed over decades while others have emerged only recently. This diversity can also be observed in the four European traditions under consideration.

The French tradition is certainly the most theoretical of these. It has three main pillars: Vergnaud's theory of conceptual fields (see Vergnaud, 1991), Brousseau's theory of didactical situations (TDS; see Brousseau, 1997) and the anthropological theory of the didactic (ATD) that emerged from Chevallard's theory of didactic transposition (see Chevallard & Sensevy, 2014). These developed over decades with the conviction that the didactics of mathematics should be a scientific field of research with fundamental and applied dimensions supported by genuine theoretical constructions and appropriate methodologies, giving an essential role to the observation and analysis of didactic systems and to didactical engineering. These theories were first conceived as tools for the understanding of mathematics teaching and learning practices and processes, taking into consideration the diversity of the conditions and constraints that shape them, and for the identification of associated phenomena, such as the 'didactic contract'. The three theories are also characterised by a strong epistemological sensitivity. Over the years, this theoretical landscape has been continuously enriched by new constructions and approaches, but efforts have always been made to maintain its global coherence.

The Dutch tradition is less diversified, as it has developed around a single approach known today as Realistic Mathematics Education (RME; see Van den Heuvel-Panhuizen & Drijvers, 2014). It also emerged in the 1970s with Freudenthal's intention to give the didactics of mathematics a scientific basis. Similar to the French case, this construction was supported by a deep epistemological reflection: Freudenthal's didactical phenomenology of mathematical structures (see Freudenthal, 1983). In this tradition, theoretical development and design are highly interdependent. This is visible in the RME structure, which is made of six principles clearly connected to design: activity, reality, level, intertwinement, interactivity and guidance. Through design research in line with these principles, many local instruction theories focusing on specific mathematical topics have been produced. RME is still in conceptual development, benefiting from interactions with other approaches such as socioconstructivism, instrumentation theory and embodied cognition theory.

In the Italian tradition, it is not equally possible to identify major theories that would have similarly emerged and developed, despite a long-term tradition of action research collaboratively carried out by mathematicians interested in education and by teachers. Progressively, however, a specific research trend has emerged from this action research and consolidated within a paradigm of research for innovation, leading to the development of specific theoretical frames and constructs (for an overview, see Arzarello & Bartolini Bussi, 1998). Boero's construct of field of experience, Bartolini Bussi and Mariotti's theory of semiotic mediation, and Arzarello's constructs of semiotic bundle and action, production and communication (APC) space represent this trend well.

In Germany, scholars since the early 1970s have aimed to create the field of didactics of mathematics as a scientific discipline, as shown by articles published in *ZDM* in 1974–75 (see Griesel, 1974; Winter, 1975; Wittmann, 1974) and the efforts made by Hans-Georg Steiner to establish an international debate on the theory of mathematics education and the underlying philosophies and epistemologies of mathematics within an international Theory of Mathematics Education (TME) group he founded in 1984. However, it would be difficult to identify a specific German way of approaching theoretical issues in the didactics of mathematics even though, seen from the outside, the interactionist approach initiated by Heinrich Bauersfeld, for example, seems to have been influential at an international level. Research in Germany currently uses a large variety of 'local' theories and corresponding research methods (for more information, see Jahnke et al., 2017).

Thus, the theoretical landscape offered by these four traditions is diverse and heterogeneous. Considering that such diversity is inherent to this field of research, the European community of research in the didactics of mathematics has developed specific efforts to build connections, an enterprise today known as 'networking between theories' (see Bikner-Ahsbahs & Prediger, 2014). Not surprisingly, researchers from these four traditions are particularly active in this area.

1.4   The Role of Design Activities for Teaching and Learning Environments

Design activities in the didactics of mathematics can involve the design of tasks, lessons, teaching sequences, textbooks, curricula, assessments, and ICT-based material or programs for teacher education and can be done by teachers, educators, textbook authors, curriculum and assessment developers, ICT designers, and researchers. Such activities can be ad hoc or research based. Without design, no education is possible. It is through designed instructional material and processes, in which the intended *what* and *how* of teaching is operationalised, that learning environments for students can be created. As such, educational design forms a meeting point of theory and practice through which they influence each other reciprocally. All four European didactic traditions reflect this role of design.

In France, the design of mathematical tasks, situations and sequences of situations is essential to didactic research and is controlled by the theoretical frameworks underlying this research (see Sect. 1.3). This is clearly reflected in the methodology of didactical engineering within the theory of didactical situations that emerged in the early 1980s. Designs are grounded in epistemological analyses, and situations are sought that capture the epistemological essence of the mathematics to be learned. In the last decade, the anthropological theory of the didactic has developed its own design perspective that gives particular importance to identifying issues that question the world and have strong mathematical potential. Design as a development activity takes place mostly within the IREMs. Dissemination happens through the publications of these institutes, professional journals, curricular resources and some textbooks. Up to now, only a few research projects were aimed at upscaling.

In the Netherlands, a strong tradition in design can be found. Making things work, looking for pragmatic solutions, creativity and innovation are typical features of the Dutch culture. This emphasis on design can also be found in the didactics of mathematics. At the end of the 1960s, the reform of mathematics education started with designing an alternative for the mechanistic mathematics education that then prevailed. Initial design activities were practice oriented. The theory development that resulted in Realistic Mathematics Education (see Sect. 1.3) grew from this practical work and later guided further design activities. Design implementation, including contexts, didactical models, longitudinal teaching-learning trajectories, textbook series, examination programs, mathematics events, and digital tools and environments, has been realised through a strong infrastructure of conferences, journals and networks.

In Italy, the role of design has also changed over time. The period from the mid-1960s to the mid-1980s was characterized by a deep epistemological concern and a strong pragmatic interest in improving classroom mathematics teaching. Theoretical reflection on didactical suggestions and their effectiveness was not so strong. The focus was on the content and its well-crafted presentation in practice, based on conceptual analyses. The period from the mid-1980s to the present can be characterised by long and complex processes targeting the development of theoretical constructs based on teaching experiments, with the design of teaching and learning environments as both an objective and a means of the experimentation.

Within the German didactic tradition, two periods can be distinguished. Before the 1970s and 1980s, design activities were mostly meant for developing learning and teaching environments for direct use in mathematics instruction. These design activities belonged to the long German tradition of *Stoffdidaktik*, which focused strongly on mathematical content and course development, with less attention on course evaluation. In the 1970s, an empirical turn occurred, resulting in design activities done to study the effect of specified didactical variables through classroom experiments. Course development became less prominent, but this was—in one strand of German didactics of mathematics—counterbalanced by defining didactics of mathematics as a 'design science' with a strong focus on mathematics. Currently, both approaches to design activities can be found in Germany and have evolved into a topic-specific didactical design research connecting design and empirical research.

1.5   The Role of Empirical Research

As discussed in Sect. 1.4, designing learning environments for mathematics has been an important activity in all four countries. This created the need to legitimise such environments. One way to do this has been to show the effectiveness of these environments by means of empirical research (whatever 'effectiveness' may mean here). Thus, with various institutional settings and with varying visibility, empirical research has an important role in the didactics of mathematics. Because of the complexity of the field, direct cause-effect research (mimicking classical natural science research) was soon found difficult, if not impossible. Nevertheless, partly as a fall-out from the need to design learning environments, empirical research in European didactics of mathematics developed a variety of questions, aims, topics and research methods such as statistical analysis with the help of tests and questionnaires, content analysis of curricula and textbooks, and classroom analysis with the help of video and observation sheets that was sometimes followed by transcript analysis (often with concepts from linguistics). More recently, triangulation and mixed methods complement the range of research methods used in empirical research in all four countries.

A major division in the plethora of empirical research is the difference between large-scale research and small and medium-sized case studies. The COACTIV study in Germany is a prototype of large-scale research. It was designed to investigate teacher competence as a key determinant of instructional quality in mathematics (for more details on this study, see Kunter et al., 2013). A contrasting example is Mithalal's case study on 3D geometry. Using Duval's *déconstruction dimensionelle* and the theory of didactical situations as the theoretical framework (see Sect. 1.3), the study took a qualitative approach to analysing students dealing with the reconstruction of a drawing showing a 3D configuration (for details see https://hal.archives-ouvertes. fr/tel-00590941).

Large-scale research can be further distinguished from medium- or small-scale research along the following lines: Large-scale studies tend to make differences within a representative sample an argument, while small- or medium-scale studies tend to make specialities of the 'case' an argument. In addition to this, empirical research can be distinguished along methodological lines: Quantitative studies tend to use sophisticated statistical techniques, while qualitative studies tend to use techniques from content analysis. In addition, there are, to an increasing proportion, mixed methods studies which use both qualitative and quantitative techniques.

If we look into the purposes of empirical research, we find commonalities and differences in these four countries. Prescriptive studies, which tend to show how things *should* be, are found in all countries, as are descriptive studies, which tend to give the best possible description and understanding of the domain under study while not being primarily interested in changing the domain. We find experimental studies on theories on the didactics of mathematics, which are undertaken to develop or elaborate a theory and put it to a test, in Italy, France and the Netherlands (less frequently in Germany), while illustrations of an existing theory (as a sort of 'existence proof') can be found in all four countries.

Another distinction is action research as opposed to fundamental research. Action research is deeply involved with the phenomena and persons under study and has the main aim of improving the actual teaching and learning. This is widespread in Italy and the Netherlands. In contrast to this, fundamental research tends to prioritise understanding of the phenomena under study and has the major aim of improving theoretical concepts: This type of research can be found in all four countries. An additional purpose of empirical research can be specific political interests (in contrast to the development of science or in addition to an interest in scientific progress and curriculum development). This type of research can be found particularly in Germany.

1.6 The Presented Cases

In Chaps. 2–7, the four didactic traditions that were referred to in the preceding sections as well as two more traditions are presented in detail.

The first presented tradition in Chap. 2 is the French. First, the emergence and development of this tradition according to the four key features are described. In particular, the three main theoretical pillars of this tradition are discussed in detail, namely the theory of didactical situations, the theory of conceptual fields and the anthropological theory of the didactic. The French tradition is then illustrated through two case studies devoted to research carried out within this tradition on line symmetry and reflection and on algebra. In the following sections, the influence of the French tradition on the mathematics education community at large is shown through the contributions of four researchers from Germany, Italy, Mexico and Tunisia. The German view of the French didactic tradition is examined through a detailed look at didactic research on validation and proof. Interactions in didactics of mathematics between France and Italy are exemplified by means of collaborative projects such as SFIDA and the personal trajectory of an Italian researcher. Didactic connections between France and Latin America are illustrated by a case from Mexico, and the long-standing connections between France and (in particular Francophone) Africa by a case from Tunisia.

In Chap. 3, the Dutch didactic tradition is presented. In the first section, the development of this tradition since the beginning of the 19th century is described with reference to the four key features. The most important feature, characteristic of the Dutch tradition, is the emphasis on design activities, strongly influenced by the IOWO institute (which since 1991 has been called the Freudenthal Institute after its first director). The second section is devoted to Adri Treffers' ideas and conceptions for RME at the primary level, with an emphasis on pupils' own productions and constructions. The third section describes the contribution of another exponent of RME, Jan de Lange, and gives examples of the use of real-world contexts for introducing and developing mathematical concepts. The fourth section gives an illustration of how the principles of RME can guide the design of a new task, using the context of a car trip to Hamburg. The chapter finishes by letting voices from abroad speak about the influence of Dutch conceptions, especially RME, on mathematics education in other parts of the world: the US, Indonesia, England, South Africa and Belgium.

The Italian didactic tradition is the content of Chap. 4. After a short historic overview, some of the crucial features that shaped Italian didactics of mathematics are presented. It becomes clear how local conditions, especially the high degree of freedom left to the teacher, influenced the design and the implementation of didactic interventions. Afterwards, the long-standing fruitful collaboration between French and Italian researchers is described and illustrated both by joint seminars and by concrete Ph.D. cases. In the final section, the collaboration between Italian and Chinese mathematics educators gives rise to a view on the Italian tradition, both culturally and institutionally, from an East Asian perspective.

Chapter 5 contains a summary of the German-language didactic tradition. The chapter starts with a historical sketch of German-speaking didactics of mathematics, starting with the 1960s. Section 5.2 gives an overview of the present situation of didactics of mathematics in the 21st century, distinguishing between three major strands: one mainly oriented towards the analysis of subject matter and the design of learning environments, one emphasising small-scale classroom studies, and one following a large-scale paradigm for analysing the learning and teaching of mathematics and the professional competencies of teachers. The last two strands are illustrated by two examples of research studies. Section 5.3 looks briefly into the future of German-language didactics of mathematics. Section 5.4 introduces perspectives from outside the German-language community. Researchers from the Nordic countries of Norway and Sweden, from Poland and from the Czech Republic present and comment on interactions concerning research in didactics of mathematics in their own countries and the German-speaking countries.

Chapter 6 presents a survey of the development of didactics of mathematics as a research domain in the Scandinavian countries of Denmark, Norway and Sweden. After an introduction, Sect. 6.2 presents a historical overview about the situation in these three countries, starting in the 1960s. In Sect. 6.3, some important trends that have been of particular importance in each of these countries are described. The following three sections take a detailed look into the situation in each of the three countries, emphasising special features of research in mathematics education in these countries such as work on mathematical competencies in Denmark, the social-cultural tradition in Norway and studies with a focus on low achievers in Sweden. Section 6.7 reports on common activities and collaborative projects, such as the NORMA conferences and the journal *NOMAD*, between these three countries and between these countries and Finland, Iceland and the Baltic countries.

The Czech and Slovak didactic tradition is discussed Chap. 7, presenting the emergence of research in didactics of mathematics in the former Czechoslovakia and the developments after the revolution in 1989. Section 7.2 shows that before 1989, Czechoslovak research developed relatively independently from other parts of Europe, yet its character shows similar key features as identified in Western European countries. In Sect. 7.3, research after 1989 is presented. The research fields are divided into four major strands: development of theories (such as the theory of generic models), knowledge and education of mathematics teachers, classroom research (e.g., related to scheme-based education at the primary level) and pupils' reasoning in mathematics. Each strand is illustrated by relevant work by Czech and Slovak researchers, with a focus on empirical research. The chapter is rounded off by naming some perspectives and challenges for didactics of mathematics in the Czech Republic and Slovakia.

References

Arzarello, F., & Bartolini Bussi, M. G. (1998). Italian trends of research in mathematics education: A national case study in the international perspective. In J. Kilpatrick & A. Sierpinska (Eds.), *Mathematics education as a research domain: A search for identity* (pp. 243–262). Dordrecht: Kluwer.

Bikner-Ahsbahs, A., & Prediger, S. (Eds.). (2014). *Networking of theories as a research practice in mathematics education.* Heidelberg: Springer.

Blum, W., Artigue, M., Mariotti, M. A., Sträßer, R., & Van den Heuvel-Panhuizen, M. (2017). European didactic traditions in mathematics: Aspects and examples from four selected cases. In G. Kaiser (Ed.), *Proceedings of the 13th International Congress on Mathematical Education—ICME-13* (pp. 291–304). Cham: Springer.

Brousseau, G. (1997). *Theory of didactical situations in mathematics.* Dordrecht: Kluwer.

Chevallard, Y., & Sensevy, G. (2014). Anthropological approaches in mathematics education, French perspectives. In S. Lerman (Ed.), *Encyclopedia of mathematics education* (pp. 38–43). New York: Springer.

Freudenthal, H. (1983). *Didactical phenomenology of mathematical structures.* Dordrecht: Kluwer.

Griesel, H. (1974). Überlegungen zur Didaktik der Mathematik als Wissenschaft. *Zentralblatt für Didaktik der Mathematik, 6*(3), 115–119.

Jahnke, H.-N., Biehler, R., Bikner-Ahsbahs, A., Gellert, U., Greefrath, G., Hefendehl-Hebeker, L., … Vorhölter, K. (2017). German-speaking traditions in mathematics education research. In G. Kaiser (Ed.), *Proceedings of the 13th International Congress on Mathematical Education—ICME-13* (pp. 305–319). Cham: Springer.

Kunter, M., Baumert, J., Blum, W., Klusmann, U., Krauss, S., & Neubrand, M. (Eds.). (2013). *Cognitive activation in the mathematics classroom and professional competence of teachers—Results from the COACTIV project.* New York: Springer.

Tobies, R. (1981). *Felix Klein.* Leipzig: Teubner.

Van den Heuvel-Panhuizen, M., & Drijvers, P. (2014). Realistic mathematics education. In S. Lerman (Ed.), *Encyclopedia of mathematics education* (pp. 521–525). Dordrecht: Springer.

Vergnaud, G. (1991). La théorie des champs conceptuels [The theory of conceptual fields]. *Recherches en Didactique des Mathématiques, 10*(2–3), 133–170.

Winter, H. (1975). Allgemeine Lernziele für den Mathematikunterricht. *Zentralblatt für Didaktik der Mathematik, 7*(3), 106–116.

Wittmann, E. (1974). Didaktik der Mathematik als Ingenieurwissenschaft. *Zentralblatt für Didaktik der Mathematik, 6*(3), 119–121.

Chapter 2
The French Didactic Tradition in Mathematics

Michèle Artigue, Marianna Bosch, Hamid Chaachoua, Faïza Chellougui, Aurélie Chesnais, Viviane Durand-Guerrier, Christine Knipping, Michela Maschietto, Avenilde Romo-Vázquez and Luc Trouche

Abstract This chapter presents the French didactic tradition. It first describes the emergence and development of this tradition according to four key features (role of mathematics and mathematicians, role of theories, role of design of teaching and learning environments, and role of empirical research), and illustrates it through

M. Artigue (✉)
University Paris-Diderot, Paris, France
e-mail: michele.artigue@univ-paris-diderot.fr

M. Bosch
University Ramon Llull, Barcelona, Spain
e-mail: mariannabosch@iqs.edu

H. Chaachoua
University of Grenoble, Grenoble, France
e-mail: hamid.chaachoua@imag.fr

F. Chellougui
University of Carthage, Tunis, Tunisia
e-mail: chellouguifaiza@yahoo.fr

A. Chesnais · V. Durand-Guerrier
University of Montpellier, Montpellier, France
e-mail: aurelie.chesnais@umontpellier.fr

V. Durand-Guerrier
e-mail: viviane.durand-guerrier@umontpellier.fr

C. Knipping
University of Bremen, Bremen, Germany
e-mail: knipping@math.uni-bremen.de

M. Maschietto
University of Modena e Reggio Emilia, Modena, Italy
e-mail: michela.maschietto@unimore.it

A. Romo-Vázquez
Instituto Politécnico Nacional, Mexico City, Mexico
e-mail: avenilderv@yahoo.com.mx

L. Trouche
IFé-ENS of Lyon, Lyon, France

© The Author(s) 2019
W. Blum et al. (eds.), *European Traditions in Didactics of Mathematics*, ICME-13 Monographs,
https://doi.org/10.1007/978-3-030-05514-1_2

two case studies respectively devoted to research carried out within this tradition on algebra and on line symmetry-reflection. It then questions the influence of this tradition through the contributions of four researchers from Germany, Italy, Mexico and Tunisia, before ending with a short epilogue.

Keywords French didactics · Mathematics education · Anthropological theory of the didactic · Theory of didactical situations · Theory of conceptual fields · Didactic research on line symmetry · Didactic research on algebra · Didactic research on proof · Didactic interactions · Influence of French didactics

This chapter is devoted to the French didactic tradition. Reflecting the structure and content of the presentation of this tradition during the Thematic Afternoon at ICME-13, it is structured into two main parts. The first part, the three first sections, describes the emergence and evolution of this tradition, according to the four key features selected to structure the presentation and comparison of the didactic traditions in France, Germany, Italy and The Netherlands at the congress, and illustrates these through two case studies. These focus on two mathematical themes continuously addressed by French researchers from the early eighties, geometrical transformations, more precisely line symmetry and reflection, and algebra. The second part is devoted to the influence of this tradition on other educational cultures, and the connections established, in Europe and beyond. It includes four sections authored by researchers from Germany, Italy, Mexico and Tunisia with first-hand experience of these interactions. Finally, the chapter ends with a short epilogue. Sections 2.1 and 2.8 are co-authored by Artigue and Trouche, Sect. 2.2 by Chesnais and Durand-Guerrier, Sect. 2.3 by Bosch and Chaachoua, Sect. 2.4 by Knipping, Sect. 2.5 by Maschietto, Sect. 2.6 by Romo-Vázquez and Sect. 2.7 by Chellougui.

2.1 The Emergence and Development of the French Didactic Tradition

As announced, we pay specific attention to the four key features that structured the presentation of the different traditions at ICME-13: role of mathematics and mathematicians, of theories, of design of teaching and learning environments, and of empirical research.

e-mail: luc.trouche@ens-lyon.fr

2.1.1 A Tradition with Close Relationship to Mathematics

Mathematics is at the core of the French didactic tradition, and many factors contribute to this situation. One of these is the tradition of engagement of French mathematicians in educational issues. As explained in Gispert (2014), this engagement was visible already at the time of the French revolution. The mathematician Condorcet presided over the Committee of Public Instruction, and well-known mathematicians, such as Condorcet, Lagrange, Laplace, Monge, and Legendre, tried to respond to the demand made to mathematicians to become interested in the mathematics education of young people. The role of mathematicians was also prominent at the turn of the twentieth century, in the 1902 reform and the emergence of the idea of *scientific humanities*. Darboux chaired the commission for syllabus revision and mathematicians strongly supported the reform movement, producing books, piloting textbook collections, and giving lectures such as the famous lectures by Borel and Poincaré. Mathematicians were also engaged in the next big curricular reform, that of the New Math period. Lichnerowicz led the commission in charge of the reform. Mathematicians also contributed through the writing of books (see the famous books by Choquet (1964) and Dieudonné (1964) offering contrasting visions on the teaching of geometry), or the organization of courses for teachers, as the APMEP[1] courses by Revuz. Today mathematicians are still active in educational issues, individually with an increasing participation in popularization mathematics activities (see the activities of the association *Animath*[2] or the website *Images des Mathématiques*[3] from the National Centre for Scientific Research), and also through their academic societies as evidenced by the role played nationally by the CFEM,[4] the French sub-commission of ICMI, of which these societies are active members.

The Institutes of Research on Mathematics Teaching (IREMs[5]) constitute another influential factor (Trouche, 2016). The creation of the IREMs was a recurrent demand from the APMEP that succeeded finally thanks to the events of May 1968. Independent from, but close to mathematics departments, these university structures welcome university mathematicians, teachers, teacher educators, didacticians and historians of mathematics who collaboratively work part-time in thematic groups, developing action-research, teacher training sessions based on their activities and producing material for teaching and teacher education. This structure has strongly influenced the development of French didactic research, nurtured institutional and scientific relationships between didacticians and mathematicians (most IREM directors were and

---

[1] APMEP: Association des professeurs de mathématiques de l'enseignement public. For an APMEP history, see (Barbazo, 2010).

[2] http://www.animath.fr (accessed 2018/01/08).

[3] http://images.math.cnrs.fr (accessed 2018/01/08).

[4] CFEM (Commission française pour l'enseignement des mathématiques) http://www.cfem.asso.fr (accessed 2018/01/08).

[5] IREM (Institut de recherche sur l'enseignement des mathématiques) http://www.univ-irem.fr (accessed 2018/01/08).

are still today university mathematicians). It has also supported the strong sensitivity of the French didactic community to epistemological and historical issues.

With some notable exceptions such as Vergnaud, the first generation of French didacticians was made of academics recruited as mathematicians by mathematics departments and working part time in an IREM. Didactic research found a natural habitat there, close to the terrain of primary and secondary education, and to mathematics departments. Within less than one decade, it built solid institutional foundations. The first doctorate programs were created in 1975 in Bordeaux, Paris and Strasbourg. A few years later, the *National seminar of didactics of mathematics* was set up with three sessions per year. In 1980, the journal *Recherches en Didactique des Mathématiques* and the biennial *Summer school of didactics of mathematics* were simultaneously created. Later on, in 1992, the creation of the ARDM[6] complemented this institutionalization process.

2.1.2   A Tradition Based on Three Main Theoretical Pillars

The didactics of mathematics emerged in France with the aim of building a genuine field of scientific research and not just a field of application for other scientific fields such as mathematics or psychology. Thus it required both fundamental and applied dimensions, and needed specific theories and methodologies. Drawing lessons from the innovative activism of the New Math period with the disillusions it had generated, French didacticians gave priority to understanding the complex interaction between mathematics learning and teaching in didactic systems. Building solid theoretical foundations for this new field in tight interaction with empirical research was an essential step. Theories were thus and are still conceived of first as tools for the understanding of mathematics teaching and learning practices and processes, and for the identification of didactic phenomena. It is usual to say that French didactics has three main theoretical pillars: the *theory of didactical situations* due to Brousseau, the *theory of conceptual fields* due to Vergnaud, and *the anthropological theory of the didactic* that emerged from the *theory of didactic transposition*, due to Chevallard. These theories are complex objects that have been developed and consolidated over decades. In what follows, we focus on a few main characteristics of each of them. More information is accessible on the ARDM and CFEM websites, particularly video recorded interviews with these three researchers.[7]

---

[6] ARDM (Association pour la recherche en didactique des mathématiques) http://ardm.eu/ (accessed 2018/01/08).

[7] The ARDM website proposed three notes dedicated to these three pioneers: Guy Brousseau, Gérard Vergnaud and Yves Chevallard. On the CFEM website, the reader can access long video recorded interviews with Brousseau, Vergnaud and Chevallard http://www.cfem.asso.fr/cfem/ICME-13-didactique-francaise (accessed 2018/01/08).

2.1.2.1 The Theory of Didactical Situations (TDS)

As explained by Brousseau in the long interview prepared for ICME-13 (see also Brousseau, Brousseau, & Warfield, 2014, Chap. 4), in the sixties, he was an elementary teacher interested in the New Math ideas and having himself developed innovative practices. However, he feared the deviations that the implementation of these ideas by elementary teachers without adequate preparation might generate. Brousseau discussed this point with Lichnerowicz (see Sect. 2.1.1) who proposed that he investigate "the limiting conditions for an experiment in the pedagogy of mathematics". This was the beginning of the story. Brousseau conceived this investigation as the development of what he called an *experimental epistemology* to make clear the difference with Piagetian cognitive epistemology. According to him, this required "to make experiments in classrooms, understand what happens…the conditions of realizations, the effect of decisions". From that emerged the 'revolutionary idea' at the time that the central object of didactic research should be the situation and not the learner, situations being conceived as a system of interactions between three poles: students, teacher and mathematical knowledge.

The theory was thus developed with the conviction that the new didactic field should be supported by methodologies giving an essential role to the design of situations able to make mathematical knowledge emerge from students' interactions with an appropriate milieu in the social context of classrooms, and to the observation and analysis of classroom implementations. This vision found its expression in the COREM[8] associated with the elementary school Michelet, which was created in 1972 and would accompany the development of TDS during 25 years, and also in the development of an original design-based methodology named *didactical engineering* (see Sect. 2.1.4) that would rapidly become the privileged methodology in TDS empirical research.

As explained by Brousseau in the same interview, the development of the theory was also fostered by the creation of the doctorate program in Bordeaux in 1975 and the resulting necessity to build a specific didactic discourse. The core concepts of the theory,[9] those of *adidactical* and *didactical situations*, of *milieu* and *didactic contract*, of *devolution* and *institutionalisation*, the three *dialectics of action, formulation* and *validation* modelling the different functionalities of mathematics knowledge, and the fundamental distinction made in the theory between "*connaissance*" and "*savoir*" with no equivalent in English,[10] were thus firmly established already in the eighties. From that time, the theory has been evolving for instance with the introduction of the hierarchy of milieus or the refinement of the concept of didactic contract, thanks to the contribution of many researchers. Retrospectively, there is no doubt that the use

---

[8]COREM: Centre pour l'observation et la recherche sur l'enseignement des mathématiques.

[9]In her text *Invitation to Didactique,* Warfield provides an accessible introduction to these concepts complemented by a glossary: https://sites.math.washington.edu/~warfield/Inv%20to%20Did66%207-22-06.pdf (accessed 2018/01/08).

[10]"Connaissance" labels knowledge engaged by students in a situation while "savoir" labels knowledge as an institutional object.

of TDS in the analysis of the functioning of ordinary classrooms from the nineties has played an important role in this evolution.

2.1.2.2 The Theory of Conceptual Fields (TCF)

Vergnaud's trajectory, as explained in the interview prepared for ICME-13, was quite atypical: beginning as a student in a school of economics, he developed an interest in theatre, and more particularly for mime. His interest for understanding the gestures supporting/expressing human activity led him to study psychology at Paris Sorbonne University, where the first course he attended was given by Piaget! This meeting was the source of his interest for analyzing the competencies of a subject performing a given task. Unlike Piaget however, he gave a greater importance to the *content* to be learnt or taught than to the *logic* of the learning. Due to the knowledge he had acquired during his initial studies, Vergnaud chose mathematical learning as his field of research; then he met Brousseau, and engaged in the emerging French community of didactics of mathematics.

In the same interview prepared for ICME-13, he emphasizes a divergence with Brousseau's theoretical approach regarding the concept of situation: while Brousseau is interested in identifying one *fundamental situation* capturing the epistemological essence of a given concept to organize its learning, Vergnaud conceives the process of learning throughout "the confrontation of a subject with a larger and more differentiated set of situations". This point of view led to the development of the *theory of conceptual fields*, a conceptual field being "a space of problems or situations whose processing involves concepts and procedures of several types in close connection" (Vergnaud, 1981, p. 217, translated by the authors). The conceptual fields of additive and multiplicative structures he has especially investigated have become paradigmatic examples. In this perspective, the difference between "connaissance" and "savoir" fades in favour of the notion of *conceptualisation*. Conceptualisation grows up through the development of *schemes*; these are invariant organisations of activity for a class of situations. The operational invariants, *concepts-in-action*, or *theorems-in-action*, are the epistemic components of schemes; they support the subjects' activity on the way to conceptualizing mathematical objects and procedures. In fact, the theory of Vergnaud has extended its influence beyond the field of didactics of mathematics to feed other scientific fields, such as professional didactics, and more generally cognitive psychology.

2.1.2.3 The Anthropological Theory of the Didactic (ATD)

As explained by Chevallard in the interview prepared for ICME-13, the theory of didactic transposition emerged first in a presentation he gave at the Summer school of didactics of mathematics in 1980, followed by the publication of a book (Chevallard, 1985a). Questioning the common vision of taught knowledge as a simple elementarization of scholarly knowledge, this theory made researchers aware of the complexity of the processes and transformations that take place from the moment it is decided that some piece of knowledge should be taught, to the moment this piece of knowledge is actually taught in classrooms; it helped researchers to make sense of the specific conditions and results of these processes (Chevallard & Bosch, 2014). The anthropological theory of the didactic (ATD) is an extension of this theory. In the interview, Chevallard makes this clear:

> The didactic transposition contained the germs of everything that followed it [...] It showed that the shaping of objects for teaching them at a given grade could not be explained only by mathematical reasons. There were other constraints, not of a mathematical nature [...] In fact, the activity of a classroom, of a student, of a scholar is embedded in the anthropological reality. (translated by the authors)

With ATD the perspective became wider (Bosch & Gascón, 2006; Chevallard & Sensevy, 2014). Institutions and institutional relationships to knowledge became basic constructs, while emphasizing their relativity. A general model for human activities (encompassing thus mathematics and didactic practices) was developed in terms of *praxeologies*.[11] ATD research was oriented towards the identification of praxeologies, the understanding of their dynamics and their conditions of existence and evolution—their ecology. In order to analyse how praxeologies 'live' and 'die', and also how they are shaped, modified, disseminated, introduced, transposed and eliminated, different levels of institutional conditions and constraints are considered, from the level of a particular mathematical topic up to the level of a given culture, a civilization or the whole humanity (see the concept of *hierarchy of didactic codetermination*[12]). Such studies are crucial to analyse what kind of praxeologies are selected to be taught and which ones are actually taught; to access the possibilities teachers and students are given to teach and learn in different institutional settings, to understand their limitations and resources, and to envisage alternatives. During the last decade, ATD has been complemented by an original form of didactic engineering in terms of *study and research paths* (see Sect. 2.1.4). It supports Chevallard's ambition of moving mathematics education from what he calls the *paradigm of visiting works* by analogy with visiting monuments, here mathematics monuments such as the Pythagorean theorem, to the paradigm of *questioning the world* (Chevallard, 2015).

---

[11] A praxeology is a quadruplet made of types of tasks, some techniques used to solve these types of tasks, a discourse (technology) to describe, explain and justify the types of tasks and techniques, and a theory justifying the technology. Types of tasks and techniques are the practical block of the praxeology while technology and theory are its theoretical block.

[12] The hierarchy of didactic codetermination categorizes these conditions and constraints according to ten different levels: topic, theme, sector, domain, discipline, pedagogy, school, society, civilization, humanity.

2.1.3   Theoretical Evolutions

The three theories just presented are the main pillars of the French didactic tradition,[13] and they are still in a state of flux. Most French researchers are used to combining them in their theoretical research frameworks depending on their research *problématiques*. However, today the theoretical landscape in the French didactic community is not reduced to these pillars and their possible combinations. New theoretical constructions have emerged, reflecting the global evolution of the field. Increasing interest paid to teacher practices and professional development has led to *the double approach of teacher practices* (Robert & Rogalski, 2002, 2005; Vandebrouck, 2013), combining the affordances of cognitive ergonomics and didactics to approach the complexity of teacher practices. Increasing interest paid to semiotic issues is addressed by the Duval's semiotic theory in terms of *semiotic registers of representations* (Duval, 1995) highlighting the decisive role played by conversions between registers in conceptualization. Major societal changes induced by the entrance into the digital era are met by the *instrumental approach to didactics*, highlighting the importance of instrumental geneses and of their management in classrooms (Artigue, 2002), and by the extension of this approach in terms of a *documentational approach to didactics* (Gueudet & Trouche, 2009).

A common characteristic of these constructions however is that they all incorporate elements of the 'three pillars' heritage. The *theory of joint action in didactics* developed by Sensevy is strongly connected with TDS and ATD (Chevallard & Sensevy, 2014); the instrumental and documentational approaches combine ATD and TCF with cognitive ergonomics. The double approach—ergonomics and didactic of teacher practices—reorganizes this heritage within a global activity theory perspective; this heritage plays also a central role in the model of *mathematical working spaces*[14] that emerged more recently (Kuzniak, Tanguay, & Elia, 2016).

These new constructions benefit also from increasing communication between the French didactic community and other research communities in mathematics education and beyond. Communication with the field of cognitive ergonomics, facilitated by Vergnaud, is a good example. Rabardel, Vergnaud's former student, developed a theory of instrumented activity (Rabardel, 1995) just when the didactic community was looking for concepts and models to analyse new phenomena arising in the thread of digitalization, more exactly when symbolic calculators where introduced in classrooms (see Guin & Trouche, 1999). Thus, the notion of scheme of instrumented action developed in a space of conceptual permeability giving birth to the instrumental approach. As this theoretical construction responded to major

---

[13]Other constructions, for instance the tool-object dialectics by Douady (1986), have also been influential.

[14]The purpose of the MWS theory is to provide a tool for the specific study of mathematical work engaged during mathematics sessions. Mathematical work is progressively constructed, as a process of bridging the epistemological and the cognitive aspects in accordance with three different yet intertwined genetic developments, identified in the model as the *semiotic, instrumental and discursive* geneses.

concerns in mathematics education (Lagrange, Artigue, Laborde, & Trouche, 2003), soon its development became an international affair, leading to the development of new concepts, such as instrumental orchestration (Drijvers & Trouche, 2008), which allowed rethinking of the teachers' role in digital environments. The maturation of the instrumental approach, the digitalization of the information and communication supports, the development of the Internet, led to consider, beyond specific artefacts, the wide set of resources that a teacher deals with when preparing a lesson, motivating the development of the documentational approach to didactics (Gueudet, Pepin, & Trouche, 2012).

2.1.4   Relationship to Design

Due to the context in which the French didactics emerged, its epistemological foundations and an easy access to classrooms provided by the IREM network, classroom design has always been conceived of as an essential component of the research work. This situation is reflected by the early emergence of the concept of *didactical engineering* and its predominant methodological role in research for several decades (Artigue, 2014). Didactical engineering is structured into four main phases: preliminary analyses; design and a priori analysis; realization, observation and data collection; a posteriori analysis and validation. Validation is internal, based on the contrast between a priori and a posteriori analyses, not on the comparison between experimental and control groups.

As a research methodology, didactical engineering has been strongly influenced by TDS, the dominant theory when it emerged. This influence is especially visible in the preliminary analyses and the design phases. Preliminary analyses systematically include an epistemological component. In the design of tasks and situations, particular importance is attached to the search for situations which capture the epistemological essence of the mathematics to be learnt (in line with the concept of fundamental situation, Sect. 2.1.2.1); to the optimization of the potential of the milieu for students' autonomous learning (adidactic potential); to the management of devolution and institutionalization processes. Didactical engineering as a research methodology, however, has continuously developed since the early eighties. It has been used with the necessary adaptations in research at all educational levels from kindergarten to university, and also in research on teacher education. Allowing researchers to explore the potential of educational designs that cannot be observed in ordinary classrooms, it has played a particular role in the identification and study of the learning and teaching potential of digital environments. The development of Cabri-Géomètre (Laborde, 1995) has had, from this point of view, an emblematic role in the French community, and at an international level, articulating development of digital environment, development of mathematical tasks and didactical research.

In the last decade, ATD has developed its own design perspective based on what Chevallard calls the *Herbartian model*, in terms of study and research paths (SRP). In it, particular importance is given to the identification of generating questions with strong mathematics potential, and also to the dialectics at stake between inquiry and the study and criticism of existing cultural answers. This is expressed through the idea of a *dialectic between medias and milieus*. A distinction is also made between *finalized SRP* whose praxeological aim is clear, and *non-finalized SRP* corresponding to more open forms of work such as interdisciplinary projects. More globally, the evolution of knowledge and perspectives in the field has promoted more flexible and collaborative forms of didactical engineering. A deep reflection on more than thirty years of didactical engineering took place at the 2009 *Summer school of didactics of mathematics* (Margolinas et al., 2011).

Design as a development activity has naturally taken place, especially in the action research activities developed in the IREMs and at the INRP (National Institute for Pedagogical Research, now IFé[15] French Institute of Education). These have been more or less influenced by research products, which have also disseminated through textbooks and curricular documents, but with a great deal of reduction and distortion. However, up to recently, issues of development, dissemination or up-scaling have been the focus of only a few research projects. The project ACE (Arithmetic and understanding at primary school)[16] supported by IFé, and based on the idea of *cooperative didactical engineering* (Joffredo-Le Brun, Morelatto, Sensevy, & Quilio, 2017) is a notable exception. Another interesting evolution in this respect is the idea of *second-generation didactical engineering* developed by Perrin-Glorian (2011).

2.1.5 The Role of Empirical Research

Empirical research has ever played a major role in the French didactics. It takes a variety of forms. However, empirical research is mainly qualitative; large-scale studies are not so frequent, and the use of randomized samples even less. Statistical tools are used, but more for data analysis than statistical inferences. *Implicative analysis* (Gras, 1996) is an example of statistical method for data analysis which has been initiated by Gras, a French didactician, and whose users, in and beyond mathematics education, organize regular conferences.

Realizations in classrooms have always been given a strong importance in empirical research. This took place within the framework of didactical engineering during the first decades of research development as explained above. However, since the early nineties the distortion frequently observed in the dissemination of didactical engineering products led to increasing attention being paid to teachers' representations and practices. As a consequence, more importance was given to naturalistic observations in empirical methodologies. The enrolment of many didacticians in the IUFM (University institutes for teacher education) from their creation in 1991,[17] and

---

[15]http://ife.ens-lyon.fr/ife (accessed 2018/01/08).

[16]http://python.espe-bretagne.fr/ace/ (accessed 2018/01/08).

[17]IUFM became in 2013 ESPÉ (Higher schools for teacher professional development and for education, http://www.reseau-espe.fr/).

the subsequent move from the work with expert teachers in the IREMs to the work with pre-service and novice teachers, also contributed to this move.

The influence of the theory of didactic transposition and subsequently of ATD also expressed in empirical research. It led to the development of techniques for the analysis of transpositive processes, the identification of mathematics and didactic praxeologies through a diversity of sources (textbooks, curricular documents, educational material…). Empirical research is also influenced by technological evolution with the increasing use of videos, for instance in studies of teacher practices, and by increasing attention being paid to semiotic and linguistic processes requiring very detailed micro-analyses and appropriate tools (Forest, 2012). The involvement of teachers working with researchers in empirical research is one of the new issues that have been developed in French didactic research. It is central in the research studies considering collective work of groups of teachers and researchers, and working on the evolution of teacher practices; it is also central for the exploration of issues such as the documentational work of teachers mentioned above (Gueudet, Pepin, & Trouche, 2013).

The next two sections illustrate this general description through two case studies, regarding the research carried out on line symmetry-reflection and on algebra over several decades. These themes have been selected to show the importance of mathematics in the French didactic tradition, and also because they offer complementary perspectives on this tradition.

2.2   Research on Line Symmetry and Reflection in the French Didactic Tradition

Line symmetry and reflection[18] appear as a fundamental subject, both in mathematics as a science and in mathematics education. In particular, as Longo (2012) points out,

> We need to ground mathematical proofs also on geometric judgments which are no less solid than logical ones: "symmetry", for example, is at least as fundamental as the logical "modus ponens"; it features heavily in mathematical constructions and proofs. (p. 53)

Moreover, several characteristics of this subject make it an important and interesting research topic for the didactics of mathematics: line symmetry is not only a mathematical object but also an everyday notion, familiar to students, and it is also involved in many professional activities; it is taught from primary school to university; among geometric transformations, it is a core object as a generator of the group of isometries of the Euclidean plane; it has played a central role in the (French) geometry curriculum since the New Math reform. This curriculum and the subsequent ones,

---

[18]In French, the expression «symétrie axiale» refers both to the property of a figure (line/reflection symmetry) and the geometric transformation (reflection). Therefore, these two aspects of the concept are probably more intertwined in French school and research than in English speaking countries. The word "réflexion" might be used in French but only when considering negative isometries in spaces of dimension greater than 2.

indeed, obeyed the logic of a progressive introduction of geometrical transformations, the other subjects being taught with reference to transformations (Tavignot,
1993)—even if this logic faded progressively since the eighties (Chesnais, 2012).
For all these reasons, the learning and teaching of reflection and line symmetry has
been the subject of a great deal of didactic research in France from the eighties up
to the present.

From the first studies about line symmetry and reflection in the eighties,
researchers have explored and modelled students' conceptions and teachers' decisions using the theory of conceptual fields (Vergnaud, 1991, 2009) and TDS
(Brousseau, 1986, 1997), and later the *cK¢ model of knowledge*[19] (Balacheff & Margolinas, 2005; Balacheff, 2013), in relation to questions about classroom design, in
particular through the methodology of didactical engineering (see Sect. 2.1). Since the
nineties and the increasing interest in understanding how ordinary classrooms work,
however, new questions have arisen about teachers' practices. They were tackled
in particular using the double didactic-ergonomic approach mentioned in Sect. 2.1.
In current research, as the role of language in teaching and learning processes has
become a crucial subject for part of the French research community, line symmetry
and reflection once again appear as a particularly interesting subject to be investigated, in particular when using a logical analysis of language (Durand-Guerrier,
2013) or studying the relations between physical actions and verbal productions in
mathematical activity.

This example is particularly relevant for a case study in the French didactic tradition, showing the crucial role of concepts and of curriculum developments in research
on the one hand, and the progressive capitalization of research results, the evolution
of research questions and the links between theoretical and empirical aspects on the
other hand.

2.2.1   Students' Conceptions, Including Proof and Proving, and Classroom Design

The first Ph.D. thesis on the teaching and learning of line symmetry and reflection
was defended by Grenier (1988). Her study connected TCF and the methodology of
didactical engineering relying on TDS. She studied very precisely students' conceptions of line symmetry, using activities where students were asked to answer open
questions such as (Fig. 2.1).

In addition, various tasks were proposed to students of constructing symmetrical drawings and figures that were crucial in identifying erroneous conceptions and

---

[19]cK¢ (conception, knowing, concept) was developed by Balacheff to build a bridge between
mathematics education and research in educational technology. It proposes a model of learners'
conceptions inspired by TDS and TCF. In it conceptions are defined as quadruplets (P, R, L, Σ)
in which P is a set of problems, R a set of operators, L a representation system, and Σ a control
structure. As pointed out in (Balacheff, 2013) the first three elements are almost directly borrowed
from Vergnaud's model of conception as a triplet.

What do these figures have in common?	  

Fig. 2.1 Example of question to study students' conceptions (Grenier 1988, p. 105)

key didactical variables of tasks. This constituted essential results on which further research was based (see below). In her thesis, Grenier also designed and trialled a didactical engineering for the teaching of line symmetry and reflection in middle school (Grade 6). Beyond the fine grained analysis of students' conceptions and difficulties presented in the study, her research produced the important result that to communicate a teaching process to teachers, it is necessary to consider not only students' conceptions, but also teachers' representations about the mathematic knowledge at stake, the previous knowledge of the students, and the way students develop their knowledge.

Relying on the results of Grenier, Tahri (1993) developed a theoretical model of students' conceptions of line symmetry. She proposed a modelling of didactical interaction in a hybrid tutorial based on the micro-world Cabri-Géomètre in order to analyse teachers' decisions. Tahri's results served then as a starting point for the research of Lima (2006). Lima (2006) referred to the cK¢ (conception, knowing, concept) model of knowledge in order to identify a priori the controls that the students can mobilize when solving problems related to the construction and the recognition of symmetrical figures. In the cK¢ model, a conception is characterized by a set of problems (P), a set of operators (R), a representation system (L), and also a control structure (S), as explained above. Lima's study showed the relevance of the modelling of the structure of control in the case of line symmetry; in particular, it allowed the author to reconstruct some coherent reasoning in cases where students' answers seemed confused. This refinement of the identification of students' conceptions of line symmetry was then used to study teachers' decisions when designing tasks aimed at allowing students to reach adequate conceptions of line symmetry. Lima concluded that a next step was to identify problems favouring the transition from a given student's conception to a target conception, both being known.

CK¢ was also used in the Ph.D. of Miyakawa (2005) with a focus on validation. Considering with Balacheff that «learning proof is learning mathematics», Miyakawa studied the relationships between mathematical knowledge and proof in the case of line symmetry. His main research question concerned the gap between pragmatic validation, using pragmatic rules that cannot be proved in the theory (e.g. relying on perception, drawings or mental experience) and theoretical validation relying on rules that can be proved in the theory (Euclidian geometry). Miyakawa especially focused on rules that grade 9 students are supposed to mobilize when asked to solve either construction problems or proving problems. He showed that, while the rules at stake were apparently the same from a theoretical point of view, students able to solve construction problems using explicitly the appropriate rules might not be able to use the corresponding rules in proving problems. The author concluded that although construction problems are playing an important role to overcome the gap between pragmatic and theoretical validation, they are not sufficient for this purpose.

A new thesis, articulating the TCF and the geometrical working spaces approach mentioned in Sect. 2.1 was then defended by Bulf (2008). She studied the effects of reflection on the conceptualization of other isometries and on the nature of geometrical work at secondary school. As part of her research, she was interested in the role of symmetry in the work of stone carvers and carpenters. Considering with Vergnaud that action plays a crucial role in conceptualization in mathematics, she tried to identify invariants through observations and interviews. The results of her study support the claim that the concept of symmetry organizes the action of these stone carvers and carpenters.

2.2.2 The Study of Teachers' Practices and Their Effects on Students' Learning

As Chevallard (1997) points out, it took a long time to problematize teaching practices:

> Considered only in terms of his weaknesses, [...] or, on the other hand, as the didactician's double [...], the teacher largely remained a virtual object in our research field. As a result, didactical modelling of the teacher's position [...] is still in its infancy. (p. 24, translated by the authors)

Results of research in didactics in France—and, seemingly, in the rest of the world—progressively led researchers to problematize the question of the role of the teacher in the teaching and learning process. Questions arose in particular from the "transmission" of didactical engineering, as illustrated by Grenier's Ph.D. thesis. For example, teachers would not offer sufficient time for the students to explore problems. Instead they would give hints or answers quickly. It also appeared that when implementing a situation, teachers would not necessarily institutionalize with students what the researcher had planned. Along the same lines, Perrin-Glorian, when working with students in schools from disadvantaged areas (Perrin-Glorian, 1992), identified that the teachers' decisions are not necessarily coherent with the logic of the situation (and knowledge at stake). Some hypotheses emerged about the factors that might cause these "distortions". Mainly, researchers explained them by the need for teachers to adjust to the reality of classrooms and students:

> But it also appears that control of problem-situations cannot guarantee the reproducibility of the process, because some of these discrepancies result from decisions that the teacher takes to respond to the reality of the class. (Grenier, 1988, p. 7, translated by the authors)

A second important hypothesis was that the researchers and teachers had different conceptions of what learning and teaching mathematics is. Hence, in the eighties, many researchers decided to investigate the role of the teacher in the teaching process and to study ordinary teachers' practices. This led to new developments within existing theories and to the emergence of a new approach. A *structured model of milieu* was developed in TDS (Brousseau, 1986; Margolinas, 2004); modelling of the teacher's position was initiated in ATD by Chevallard (1999). Robert and Rogalski developed the double approach of teachers' practices mentioned in 1.3 based on Activity Theory and the socio-historical theory of Vygostky, and connecting didactic and ergonomic points of view on teachers' practices. Research in this approach is driven by the investigation of regularities and variability of ordinary teaching practices depending on contexts, mathematical subjects, grades, teachers, etc. This investigation then allows the identification of causes and rationales underlying teachers' practices. Another focus is to investigate the effects of teaching practices on students' learning. Chesnais' Ph.D. thesis about the teaching and learning of reflection and line symmetry (Chesnais, 2009) is a good example of this evolution. Relying on previous research and using the didactic and ergonomic double approach, she compared the teaching and learning of these subjects in the 6th grade between a school situated in a socially disadvantaged area and an ordinary one. The research was based on both 'naturalistic observation' (during the first year) and an experiment (during the second year) which consisted of the transmission of a teaching scenario about reflection and line symmetry elaborated by one teacher from the ordinary school to another one from the socially disadvantaged school. The results showed that socially disadvantaged students could perform "as well as" ordinary ones provided that certain conditions are fulfilled. In particular, the following crucial conditions were identified: an important conceptual ambition of the teaching scenario, its coherence and "robustness" and the fact that the teacher receiving it is sufficiently aware of some specificities of the content and students' learning difficulties. Moreover, the research showed that multiple reasons drove the teachers' choices, and explained some differences identified between them: the fact that they taught to different audiences had probably a great influence but also their experience in teaching, determining their ability to identify the important issues about the teaching of line symmetry and reflection (in this case, it appeared that the experienced teacher of the ordinary school had a more coherent idea about the teaching of line symmetry and reflection because she had experienced teaching with older and more detailed teaching instructions). The research also suggested that collaborative work between teachers might be a good lever for professional development under certain conditions (in this work, the role of the researcher as an intermediary was crucial because the first teacher was not clearly conscious of what made her scenario efficient).

2.2.3   Current Research

A consistent part of recent French research heads toward a thorough investigation of the role of language in the teaching and learning of mathematics. Research globally considers language either as an object of learning (as part of concepts), as a medium for learning (its role in the conceptualisation process) and for teaching, and/or as a methodological means for researchers to get access to students' and teachers' activity. Line symmetry and reflection represent once again an interesting subject with regard to the role of language. Indeed, a logical analysis (Vergnaud, 2009; Durand-Guerrier, 2013) shows that symmetry can be considered as a property of a given figure but also, via reflection, as a ternary relation involving two figures and an axis, or as a geometric transformation involving points, or even as a binary relation—when considering two figures and questioning the existence of a reflection transforming one into the other. Studying how these "variations of the meanings of words" may be expressed in the French language shows an incredible complexity, in particular because of the polysemy of the words "symétrie" and "symétrique". This makes symmetry a good subject to study the relationships between action and language in mathematical activities, and how teachers and students deal with this complexity. For example Chesnais and Mathé (2015) showed that 5th grade[20] pupils' conceptualisation of the "flipping over property"[21] of reflection results from the articulation of several types of interactions between students and milieu mediated by language, instruments like tracing paper, and by the teacher: for example, they showed that the manipulation of the tracing paper (flipping it to check if the initial figure and its image match) needs to be explicitly identified and that it is complementary to the use of an adequate vocabulary. It also appeared that teachers identified these issues differently.

Questions about teacher education and the development of teaching resources also play an important role in recent developments of research in the field. Here again, line symmetry and reflection were chosen as subjects for research. For instance, Perrin-Glorian elaborated the concept of second generation didactical engineering mentioned in Sect. 2.1 in the context of a long term and original research project regarding the teaching and learning of geometry, in which reflection and line symmetry play a crucial role (Perrin-Glorian, Mathé, & Leclercq, 2013). Searching for the construction of coherence from primary to secondary school, this project also led to a deepening of reflection on the role that the use of instruments can play in a progressive conceptualisation of geometrical objects.

We cannot enter into more details about this important set of research, but through this case study, we hope to have made clear an important feature of the development of didactic research in the French tradition: the intertwined progression of research questions, theoretical elaborations and empirical studies, coherently over long periods of time.

---

[20] The 5th grade corresponds to the last grade of primary school in France.

[21] We refer here to the fact that reflection is a *negative* isometry.

2.3   Research on School Algebra in the French Didactic Tradition. From Didactic Transposition to Instrumental Genesis

Research on school algebra started at the very beginning of the development of the French didactic tradition in the 1980s with the first studies of the didactic transposition process (Chevallard, 1985a, b, 1994). Since then, and for more than 30 years, school algebra has been the touchstone of various approaches and research methodologies, which spread in the French, Spanish and Italian speaking communities, thanks to numerous collaborations and common research seminars like SFIDA (see Sect. 2.5).

Fortunately, all this work has been synthesized in two very good resources. The first one is a special issue of the journal *Recherches en Didactique des Mathématiques* (Coulange, Drouhard, Dorier, & Robert, 2012), which presents a summary of recent works in the subject, with studies covering a wide range of school mathematics, from the last years of primary school to the university. It includes sixteen papers grouped in two sections: a first one on teaching algebra practices and a second one presenting cross-perspectives with researchers from other traditions. The second resource is the survey presented by Chaachoua (2015), and by Coppé and Grugeon (2015) in two lectures given at the 17th *Summer school of didactics of mathematics* in 2013 (see Butlen et al., 2015). They focus on the effective and potential impacts between didactic research, the teaching profession and other instances intervening in instructional processes (curriculum developers, textbooks authors, teacher educators, policy makers, etc.). The discussion also deals with transfers that have not taken place and highlights various difficulties that seem to remain embedded in the school institution.

We cannot present this amount of work in a few pages. Instead of focusing on the results and the various contributions of each team, we have selected three core research questions that have guided these investigations in the field of secondary school mathematics. First, the analysis of the didactic transposition process and the associated questions about what algebra is and what kind of algebra is taught at school. Second, research based on didactical engineering proposals which address the question of what algebra could be taught and under what kind of conditions. Finally, both issues are approached focusing on ICT to show how computer-assisted tools can modify not only the way to teach algebra but also its own nature as a mathematical activity.

2.3.1   What Algebra Is to Be Taught: Didactic Transposition Constraints

To understand the role played by research on school algebra in the development of the field of didactics of mathematics, we should look back to the 1980s and Cheval-lard's attempt to approach secondary school mathematics from the new perspective opened by the theory of didactic situations (TDS). At this period, TDS was mainly focused on pre-school and primary school mathematics. Its first enlargements to secondary school mathematics gave rise to the analyses in terms of didactic transposition (Chevallard, 1985a). Algebra was one of the case studies that received more attention. Contrary to the majority of investigations of this period, studies on the didactic transposition processes directly adopted an institutional perspective anchored in deep epistemological and historical analyses.

The first studies (Chevallard, 1985b) pointed out the nature and function of algebra in scholarly mathematics since the work of Vieta and Descartes, and its fading importance in secondary school curricula after the New Maths reform. What appears is a certain lack of definition of algebra as a school mathematical domain (in France as well as in many other countries), centred on the resolution of first and second degree equations. Many elements of what constitutes the driving force of the algebraic work (use of parameters and inter-play between parameters and unknowns, global modelling of arithmetic and geometrical systems, etc.) have disappeared from school curricula or only play a formal role. Part of this evolution can be explained by a cultural difficulty in accepting the primarily written nature of algebra, which strongly contrasts with the orality of the arithmetical world (Chevallard, 1989, 1990; see also Bosch, 2015).

A first attempt to describe the specificities of algebraic work was proposed by considering a broad notion of modelling that covers the modelling of extra-mathematical as well as mathematical systems (Chevallard, 1989). Analyses of algebra as a modelling process were later developed in terms of the new epistemological elements proposed by the Anthropological Theory of the Didactic based on the notion of praxeology. This extension brought about new research questions, such as the role of algebra and the students' difficulties in school institutional transitions (Grugeon, 1995) or the constraints appearing in the teaching of algebra when it is conceived as a process of algebraisation of mathematical praxeologies (Bolea, Bosch, & Gascón, 1999). In this context, algebra appears linked to the process of modelling and enlarging previously established praxeologies.

2.3.2   Teaching Algebra at Secondary School Level

While the aforementioned studies are more focused on what is considered as the external didactic transposition—the passage from scholarly knowledge to the knowledge to be taught—the question of "what algebra could be taught" is addressed by investigations approaching the second step of the didactic transposition process, the internal transposition, which transforms the knowledge to be taught into knowledge actually taught. This research addresses teaching and learning practices, either from a 'naturalistic' perspective considering what is effectively taught and learnt as algebra at secondary level—as well as what is not taught anymore—or in the design and implementation of new proposals following the methodology of didactical engineering. Research questions change from "What is (and what is not) algebra as knowledge to be taught?" to "What can be taught as algebra in teaching institutions today?". However, the answers elaborated to the first question remain crucial as methodological hypotheses. Experimental studies proposing new conditions to teach new kinds of algebraic activities rely on the previous elaboration of a priori epistemological models (what is considered as algebra) and didactic models (how is algebra taught and learnt).

The notion of calculation program (*programme de calcul*), used to rebuild the relationships between algebra and arithmetic in a modelling perspective, has been at the core of some of these instructional proposals, especially in the work carried out in Spain by the team led by Gascón and Bosch. Following Chevallard's proposal to consider algebra as the "science of calculation programs", a *reference epistemological model* is defined in terms of stages of the process of algebraisation (Bosch, 2015; Ruiz-Munzón, Bosch, & Gascón, 2007, 2013). This redefinition of school algebra appears to be an effective tool for the analysis of curricula and traditional teaching proposals. It shows that the algebraisation process in lower secondary school mathematical activities is very limited and contrasts with the "fully-algebraised" mathematics in higher secondary school or first year of university. It also provides grounds for new innovative instructional processes like those based on the notions of *study and research activities* and *study and research paths* (see Sect. 2.1.4) that cover the introduction of negative numbers in an algebraic context and the link of elementary algebra with functional modelling (Ruiz-Munzón, Matheron, Bosch, & Gascón, 2012).

2.3.3   Algebra and ICT

Finally, important investigations of school algebra address questions related to the integration of ICT in school mathematics. They look at the way this integration might influence not only students' and teachers' activities, but also the nature of algebraic work when paper and pencil work is enriched with new tools such as a spreadsheet, a CAS or a specific software specially designed to introduce and make sense of elementary algebraic manipulations. This research line starts from the hypothesis that ICT tools turn out to be operational when they become part of the students' adidactic milieu, thus focusing on the importance and nature of ICT tools feedback on the resolution of tasks. However, the integration of ICT tools in the adidactic milieu cannot be taken for granted as research on instrumental genesis has shown. Many investigations have produced evidence on new teachers' difficulties and problems of legitimacy to carry out such integration, pointing at didactic phenomena like the double reference (paper and pencil versus ICT tools) (Artigue, Assude, Grugeon, & Lenfant, 2001) or the emergence of new types of tasks raised by the new semiotic representations produced by ICT tools. In fact, the epistemological and didactic dimensions (what is algebra and how to teach it) appear so closely interrelated that a broad perspective is necessary to analyse the didactic transposition processes. The instrumental approach of technological integration emerged, especially in the case of the use of CAS and technologies not initially thought for teaching. This approach was then extended to other technologies and has experienced significant internationalization (Drijvers, 2013).

In respect to technologies specifically designed for teaching, particular software devices have been designed on the basis of didactic investigations. They thus appear as a paradigm of connection between fundamental research, teaching development and empirical validation of ICT didactic tools (Chaachoua, 2015). The first one is *Aplusix*,[22] a micro world especially designed for the practice of elementary algebra that remains very close to the students' paper and pencil manipulations, while providing feed-back based on a detailed epistemological and didactic analysis of potential learning processes (Trgalová & Chaachoua, 2009). Aplusix is based on a model of reasoning by equivalence between two algebraic expressions which is defined by the fact that they have the same denotation (Arzarello, Bazzini, & Chiappini, 2001). Therefore, Aplusix enables students to work on the relationship between sense and denotation, which is essential to effectuate and understand the transformation of algebraic expressions. The second device is Pépite, a computer-based environment providing diagnostics of students' competences in elementary algebra, as well as tools to manage the learning heterogeneity with the proposal of differentiating teaching paths (Pilet, Chenevotot, Grugeon, El Kechaï, & Delozanne, 2013). Both Aplusix and Pépite are the fruit of long term and continuous research work in close collaboration with computer scientists.

On the whole, all these investigations have contributed to a fundamental epistemological questioning on the nature of algebraic work and its components, sometimes enriched by linguistic or semiotic contributions. They all share the research aim of better identifying the universe of didactic possibilities offered by today's school systems and of determining the local conditions that would allow a renewed teaching of the field, directly aligned with mathematical work in primary school and connected with the world of functions and calculus for which it is preparatory

The longitudinal dimension of algebra and its predominance in higher secondary education and beyond show that one cannot approach school algebra without questioning the rationale of compulsory education as a whole and how it can supply the mathematical needs of citizens. We are thus led to the initial project of Brousseau and the fundamental problems that motivated the development of TDS: to reconstruct the compulsory mathematical curriculum based on democratic and effective principles that can be discussed in an objective and non-authoritarian way.

Let us finish this section with a tribute to Jean-Philippe Drouhard, one of the founders of the Franco-Italian Seminar in Didactics of Algebra (SFIDA, 1992–2012, see Sect. 2.5), who took special leadership in the development of the research on school algebra and devoted his research life in didactics to the study of the semio-linguistic complexity of elementary algebra (Drouhard, 2010).

After these two cases studies, we enter the second part of this chapter which is devoted to the influence of the French didactic tradition on other educational cultures, and the connections established, in Europe and beyond. These connections have a

---

[22]http://www.aplusix.com.

long history as evidenced in Artigue (2016), and the IREMs have played an important role in their development, together with the many foreign students who have prepared their doctorate in France or been involved in co-supervision programs since the early eighties. The authors of the four following sections, from Germany, Italy, Mexico and Tunisia were such doctorate students who have now pursued their careers in their own country. These four case studies illustrate different ways through which a research tradition may diffuse, influencing individual researchers or communities, and the source of enrichment that these interactions are for the tradition itself.

2.4   View of the French Tradition Through the Lens of Validation and Proof

In this section I (Christine Knipping) present my view of the French didactic tradition. I take the position of a critical friend (the role given to me at ICME-13), and I consider this tradition through the lens of didactic research on validation and proof. This topic is at the core of my own research since my doctoral work, which I pursued in France and Germany. I will structure the section around three main strengths that I see in the French tradition: Cohesion, Interchange, and Dissemination. The first strength is *Cohesion* as the French community has a shared knowledge experience and theoretical frameworks that make it possible to speak of a French Didactique. The second strength is an open *Interchange* within the community and with others. The third strength allows the *dissemination* of ideas in the wider world of mathematics education. These three strengths, which have also influenced my own research, will be illustrated by examples from a personal perspective.

2.4.1   Cohesion

As a student, having just finished my Masters Degree in Mathematics, Philosophy and Education in the 1990s in Germany, I went to Paris, and enrolled as a student in the DEA-Programme in Didactics of Mathematics at University Paris 7 (see Sect. 2.1). Courses were well structured and introduced students to key ideas in the French didactic tradition: the theory of didactical situations (TDS) developed by Brousseau, the theory of conceptual fields due to Vergnaud, and the anthropological theory of the didactic that emerged from the theory of didactic transposition, conceptualised by Chevallard. The courses were taught by a wide range of colleagues from the research group DIDIREM of Paris 7, which is now the LDAR (*Laboratoire de Didactique André Revuz*). Among others Artigue, Douady, Perrin-Glorian, as well as colleagues from several teacher training institutes (IUFM) were involved. These colleagues not only introduced us to the theoretical pillars of the French tradition in mathematics education, but they also showed us how their own research and recent doctoral work was based on these traditions. This made us aware of the power of French conceptual frameworks and demonstrated vividly how they could be applied. It also showed us some empirical results these frameworks had led to and how French research in mathematics education was expanding quickly at that time. The specific foci of research and research questions in the French community were an obvious strength, but phenomena that were not in these foci were not captured. A few of our professors reflected on this and made us aware that classroom interactions, issues of social justice and cultural contexts were more difficult to capture with the given theoretical frameworks. Also validation and proof, a topic I was very interested in, was hardly covered by our coursework in Paris, while in Grenoble there was a clear research focus on validation and proof since the 1980s (Balacheff, 1988), that probably was reflected in their DEA programme at the time.

The French National Seminar in Didactics of Mathematics, which we were invited to attend, was another experience of this phenomenon of cohesion. Many of the presentations at the National Seminar were based on the three pillars of French didactics and also French Ph.D.'s followed these lines. But there were topics and approaches beyond these frameworks that seemed to be important for the French community; validation and proof was apparently one of them. Presentations in this direction were regularly and vividly discussed at the National Seminar and quickly became public knowledge within the community. For example, Imre Lakatos' striking work *Proofs and Refutations* was not only translated by Balacheff and Laborde into French in 1984, but also presented to the French community at one of the first National Seminars. Looking through the *Actes du Séminaire National de Didactique des Mathématiques* shows that validation and proof is a consistent theme over decades. In this area of research French colleagues recognise and reference each other's work, but cohesion seems less strong in this context. Besides Balacheff's first school experiments with Lakatos' quasi-empirical approach (Balacheff, 1987), Legrand introduced the *scientific debate* as another way to establish processes of validation and proof in the mathematics class (Legrand, 2001). Coming from logic, Arsac and Durand-Guerrier approached the topic from a more traditional way, attempting to make proof accessible to students from this side (Arsac & Durand-Guerrier, 2000). Publications not only in the proceedings of the National Seminar, but also in the journal *Recherches en Didactique des Mathématiques* (RDM), made the diverse approaches in the field of validation and proof accessible. So in this context Interchange seemed the strength of the French community.

2.4.2   Interchange

Such interchange, within the community as described above but also with researchers from other countries, is in my view another striking strength. The *Colloque Franco-Allemand de Didactique des Mathématiques et de l'Informatique* is one example of this and valued by its publication in the book series associated with *Recherches en Didactique des Mathématiques* (Laborde, 1988). Many other on-going international exchanges, as described for example in the following sections of this chapter, also illustrate this strength of the French community. I received a vivid impression of the passion for interchange at the tenth *Summer school of didactics of mathematics* held in Houlgate in 1999. The primary goal of such summer schools is to serve as a working site for researchers to study the work of their colleagues. Young researchers are welcome, but not the focus. Consistently colleagues from other countries are invited. Here Boero (Italy), Duval (France) and Herbst (USA) contributed to the topic "Validation, proof and formulation", which was one of the four themes in 1999. Over the course of the summer school their different point of views—cognitive versus socio-cultural—became not only obvious, but were defended and challenged in many ways. These were inspiring debates. This interchange was also reflected in the *International Newsletter on the Teaching and Learning of Proof*,[23] established in 1997 whose first editor was Balacheff. Innovative articles and also the careful listing of recent publications made this Newsletter a rich resource for researchers in the area at the time. Colleagues from diverse countries published and also edited this online journal, which continues with Mariotti and Pedemonte as editors. Debates on argumentation and mathematical proof, their meaning and contextualization, as well as cultural differences are of interest for the Newsletter. Sekiguchi's and Miyazaki's (2000) publication on 'Argumentation and Mathematical Proof in Japan' for example already explicitly addressed this issue. One section in the Newsletter also explicitly highlights working groups and sections of international conferences that deal with the topic of validation and proof. References to authors and papers are listed and made accessible in this way. Interchange is therefore highly valued and also expressed by the fact that some articles are published in multiple languages (English, French, Spanish).

2.4.3   Dissemination

Interchange is in multiple ways related to dissemination, another strength of the French didactic tradition. Looking for vivid exchange also implies that perspectives and approaches from the community become more widely known and disseminated. The research work of Herbst is prominent in this way; French tradition had a visible impact on his own unique work, which then influenced not only his doctoral students but also other colleagues in the US. This is evident as well in the many bi-national theses (French-German; French-Italian) in mathematics education, including quite a few on the topic of validation and proof, my own Ph.D. thesis among them (Knipping, 2003). French universities were in general highly committed to this kind of double degree and had international graduate programs and inter-university coalitions with many countries. As Ph.D. students we were highly influenced by French research traditions and incorporated ideas from the French mathematics education community into our work. Pedemonte's Ph.D. thesis (2002) entitled *Etude didactique et cogni-*

---

[23] http://www.lettredelapreuve.org (accessed 2018/01/08).

tive des rapports de l'argumentation et de la démonstration dans l'apprentissage des mathématiques (Didactic and cognitive analyses of the relationship between argumentation and proof in mathematics learning) is an example of this. Her work and ideas were then further disseminated into the international validation and proof research community.

The working group on *argumentation and proof,* which has been meeting since the Third Conference of the European Society for Research in Mathematics Education (CERME 3, 2003) in Bellaria has been a vivid place not only for discussion and exchange, but also a site where ideas from the French community have continuously been prominent. French colleagues are always present, serving in guiding functions, and scholars like me who are familiar with ideas and approaches of the French tradition and have actively used them in our own work spread these ideas further into the international community. Interesting crossover work has also emerged throughout the years between different disciplines. Miyakawa, who also did his Ph.D. work in Grenoble with Balacheff (Miyakawa, 2005), is another interesting example of a colleague who stands for dissemination of French ideas and is interested in the kind of interdisciplinary work that I see as characteristic for French research in the field of validation and proof. He is now well established in Japan but he continues to collaborate with French colleagues and reaches out to other disciplines. Recently he presented at CERME 10 a paper with the title *Evolution of proof form in Japanese geometry textbooks* together with Cousin from the Lyon Institute of East Asian Studies (Cousin and Miyakawa, 2017). Collaboratively they use the anthropological theory of the didactic (ATD) to study the didactic transposition of proofs in the Japanese educational system and culture and to better understand proof taught/learnt in this institutional context. Reflecting on the conditions and constraints specific to this institution Miyakawa and Cousin help us also to see in general the nature of difficulties for students in the context of proof-and-proving from a new perspective. French theoretical frameworks are again fruitful for this kind of inter-cultural-comparative work.

In closing, from my perspective as a critical friend working in the field of validation and proof, these three strengths, Cohesion, Interchange, and Dissemination, have contributed to the success of the French Didactique. As a *critical* friend, I should also observe that each of these strengths comes with costs. Theoretical cohesion can limit the research questions that can be addressed, and research groups strongly focused on one area inevitably neglect others. This is reflected in some limitations in interchange and dissemination. French voices are clearly heard in some contexts, but hardly at all in others, and interchanges are sometimes unbalanced. Overall, however, it is clear that France has been fortunate to have a strong community in didactics, supported by a range of institutions that foster interchange and dissemination, of which this set of presentations at ICME is yet another example.

2.5 Didactic Interactions Between France and Italy. A Personal Journey

Didactic interactions between Italy and France have a long history. For instance, Italian researchers participated in the French *Summer schools of didactics of mathematics* from the first. In this section, after pointing out some structures that have nurtured these interactions, I (Michela Maschietto) present and discuss them through the lenses of my personal experience, first as a doctoral student having both French and Italian supervisors, then as an Italian researcher regularly involved in collaborative projects with French researchers. I also approach them in a more general way, considering both the cultural and institutional conditions in which the research has developed in the two countries.

2.5.1 Opportunities for Collaboration: SFIDA, Summer Schools and European Projects

Among the different institutional structures that provided opportunities for collaboration between French and Italian researchers, SFIDA[24] (Séminaire Franco-Italien de Didactique de l'Algèbre) certainly played a crucial role. The idea of this seminar arose from the interest in teaching and learning algebra shared by the researchers of the Italian teams at the University of Genova and Turin (respectively, directed by Boero and Arzarello) and the French team at the University of Nice (directed by Drouhard). SFIDA sessions were organized twice per year from 1993 to 2012 (SFIDA-38 was the last edition), alternatively by the three research teams, and held in their respective universities. A unique feature of this seminar was that everyone spoke his/her own language, as the programs of each session show. This attitude to overcome language constraints fostered the participation of researchers from other universities, and also students, both Italian and French. SFIDA was not only a place for sharing projects or work in progress, but the seminar functioned also as a working group that allowed the emerging of new ideas in this field of research, as attested by the articles devoted to this seminar in the second part of the special issue of *Recherches en Didactique des Mathématiques* on didactic research in algebra (Coulange, Drouhard, Dorier, & Robert, 2012) already mentioned in Sect. 2.3.

Other scientific events allowed the two communities to meet each other and collaborate, like the French Summer school of didactics already mentioned and the conferences of the *Espace Mathématique Francophone*. The participation and contribution of Italian researchers and teacher-researchers to those events, the involvement of Italian researchers in their scientific and organizing committees strengthened the relationships between the communities. The team of the University of Palermo (directed by Spagnolo) was for itself especially involved in the regular scientific

---

[24]https://sites.google.com/site/seminairesfida/Home/ (accessed 2018/02/10).

meetings of the group on Implicative Analysis (Gras, 1996), and even organized two of them (in 2005 and 2010). Spagnolo, supervised by Brousseau was also one of the first Italian students to get his Ph.D. in a French university (Spagnolo, 1995), together with Polo supervised by Gras (Polo Capra, 1996). Furthermore, around the years 2000, several Italian students carried out their doctoral thesis in different French universities.

French and Italian research teams were also involved in several European projects on the use of technologies, on teacher training and on theoretical perspectives. For instance, the ReMATH project (Representing Mathematics with Digital Media) has focused on the analysis of the potentiality of semiotic representations offered by dynamic digital artefacts (Kynigos & Lagrange, 2014). Adopting a perspective of networking among theoretical frameworks, this project has fostered the development of specific methodologies for such networking like the idea of cross-experiment (Artigue & Mariotti, 2014). In recent times, other research teams collaborated within the FASMED project on formative assessment.[25]

Despite those collaborations, relevant differences exist between the Italian and French traditions in mathematics education: they do not only have to deal with differences in theoretical frameworks, but also with cultural differences of the two communities in which research is carried out, as explained in Chap. 4. A critical perspective on them has been proposed by Boero, one of the promoters of SFIDA, who highlighted some difficulties to establish collaborations between French and Italian researchers since the beginning of SFIDA. For Boero (1994), they were due to:

- Italian researchers had been more interested in studying the relationships between innovative didactical proposals and their development in classes than modelling didactical phenomena, as in the French tradition;
- The experimental parts of the Italian research involving classes were not situations the researcher studied as an external observer, but they were an opportunity to make more precise and test the hypotheses about the Italian paradigm of "research for innovation".
- The presence of several teacher-researchers in Italian research teams.

By a cultural analysis of the context in which researchers work Boero deepens his reflection in a more recent contribution: he claims that he is "now convinced that these difficulties do not derive only from researchers' characteristics and personal positions, but also (and perhaps mainly) from ecological conditions under which research in mathematics education develops" (Arcavi et al., 2016, p. 26). Among these conditions, he especially points out: the features of the school systems (i.e., the Italian national guidelines for curricula are less prescriptive than French syllabuses and primary school teachers in Italy usually teach and follow the same students for five years); the economic constraints of research and the weight of the cultural environment (i.e., the cultural and social vision of mathematics, the spread of the

---

[25] Improving Progress for Lower Achievers through Formative Assessment in Science and Mathematics Education, https://research.ncl.ac.uk/fasmed/ (accessed 2018/02/10).

idea of mathematics laboratory, the development of mathematics research, and the influence of sociological studies that Boero considers weaker in Italy than in France).

I would like to give now another vision of the didactic relationships between Italy and France by reflecting on my personal scientific journey, from the perspective of boundary crossing (Akkerman & Bakker, 2011), as my transitions and interactions across different sites, and boundary objects, as artefacts, have had a bridging function for me.

2.5.2   A Personal Scientific Journey

This journey started at the University of Turin, where I obtained a one-year fellowship to study at a foreign university. The University of Bordeaux I, in particular the LADIST (*Laboratoire Aquitain de Didactique des Sciences et Techniques*) directed by Brousseau was my first destination, my first boundary crossing. At the LADIST I became more deeply involved in the Theory of Didactical Situations (Brousseau, 1997) that I had previously studied. A fundamental experience for me as a student in mathematics education was the observation of classes at the *École Michelet*, the primary school attached to the COREM (see Sect. 2.1). The activities of the COREM allowed me to compare the experimental reality with the factual components that Brousseau highlighted in his lessons to doctoral students at the LADIST. My first personal contact with French research was thus characterized by discussions, passion and enthusiasm for research and, of course, by the people I met. At the end of my fellowship, I moved to Paris to prepare a DEA.[26] There I met Artigue and other French colleagues, and I read Boero's paper (Boero, 1994) quoted above that encouraged me to become aware of the potential of my boundary crossing.

After the DEA, I continued my doctoral studies within an institutional agreement between the University of Paris 7 and the University of Turin. I had two supervisors (Artigue and Arzarello) from two didactic cultures who had not yet collaborated. Retrospectively, I can claim that the dialogue between these cultures was under my responsibility. Ante litteram, I looked for a kind of networking strategy (Prediger, Bikner-Ahsbahs, & Arzarello, 2008) appropriate to the topic of my research: the introduction of Calculus in high school with graphic and symbolic calculators (Maschietto, 2008). From the French culture I took the methodology of didactical engineering (Artigue, 2014) with its powerful a priori analysis, the idea of situation and a-didacticity, and the instrumental approach (see Sect. 2.1). From the Italian culture, I took the strong cognitive component following embodied cognition, a semiotic focus with the attention to gestures and metaphors, and the cognitive roots of concepts. The experimental part of my research was a didactical engineering carried out in some Italian classes. I had to negotiate the planned situations with the teachers of these classes who were members of the research team of the University of Turin. In that process, a relevant element was the a priori analysis of the planned situations.

---

[26]Diplôme d'Etudes Approfondies.

It was a powerful tool for sharing the grounded idea of the didactical engineering with those teachers who did not belong to the French scientific culture, and became a boundary object.

At the end of my doctoral period, I moved to the University of Modena e Reggio Emilia with a research fellowship within the European project on mathematics exhibitions *Maths Alive*. Finally, I got a research position in that university some years later, and I currently work there. In Modena, I met the framework of the Theory of Semiotic Mediation (Bartolini Bussi & Mariotti, 2008) (see Chap. 4), and the mathematical machines. These are tools mainly concerning geometry, which have been constructed by teacher-researchers (members of the team of that University directed by Bartolini Bussi, now Laboratorio delle Macchine Matematiche[27]) following the Italian tradition of using material tools inspired by the work of Castelnuovo in mathematics laboratory. In my first designs of laboratory sessions, I tried to locally integrate the TDS and the Theory of Semiotic Mediation by alternating a moment of group work with a-didactical features and a moment of collective discussion with an institutional component, with the aim of mediating mathematical meanings. Another boundary crossing for me.

The collaboration with French colleagues was renewed when I moved to the INRP[28] in Lyon for two months as visiting researcher in the EducTice team. My new position as a researcher and not as a student, changed the conditions, making more evident the potential of boundary crossing. The work with the EducTice team was the source of new insights and productive exchanges about: the notion of mathematics laboratory and the use of tools in mathematics education (Maschietto & Trouche, 2010); the notion of resources and collaborative work among teachers from the perspective of the Documentational Approach (see Sect. 2.1); and the role of digital technologies in learning, teaching and teacher education. The second and third elements contributed to the MMLAB-ER project I was involved in Italy in terms of planning and analysis of the teachers' education program (Maschietto, 2015). The first and third elements have been essential in the development of the original idea of *duo of artefacts* based on using a material artefact and a digital one in an intertwined way in the same learning project (Maschietto & Soury-Lavergne, 2013). We jointly planned the design of the digital counterpart of a material artefact,[29] the tasks with the duo of artefacts, resources and training for primary school teachers, and the idea of duo of artefacts developed as a boundary object for all of us, bridging our respective practices.

The challenge of boundary crossing is to establish a continuity across the different sites and negotiated practices with other researchers and teachers. This story tells my personal experience of boundary crossing between the French and Italian didactic cultures. This is a particular story, but not an isolated case. It illustrates how

---

[27]MMLab, www.mmlab.unimore.it (accessed 2018/02/10).

[28]Institut National de Recherche Pédagogique, now Institut Français de l'Éducation (http://ife.ens-lyon.fr/ife) (accessed 2018/02/10).

[29]In this case, the material artefact is the "pascaline Zero + 1", while the digital counterpart is the "e-pascaline" constructed in the new Cabri authoring environment.

the repeated boundary crossing of individual researchers has contributed and still contributes to the dissemination of the French didactic culture and, in return, to its enrichment by other didactic cultures.

2.6   Didactic Interactions Between France and Latin-America: The Case of Mexico

In Latin America, as in other regions of the world, Mathematics Education emerged in the 1970s, and its development was marked by a strong relationship with the French community of educational mathematicians. Initially, the Department of Mathematics Education at the Centro de Investigación y de Estudios Avanzados (DME-CINVESTAV, created in 1975), and the network of French IREMs have made exceptional contributions to this process by establishing relations between specialists in Mexico and France. Over time, other institutions emerged and the French theoretical approaches were adapted to respond to different educational needs of the continent, which resulted in a greater theoretical development: production of new tools, broadening of notions, questioning of scope, study of new issues, etc. Analyzing the case of Mexico, the focus of this section, we show other research communities how the relationships between institutions and research groups can generate scientific advances with social impact. Also, it leads to important reflections on current challenges of Latin America reality, especially that of achieving quality education for all, a key element in securing peace and social development.

2.6.1   The DME at CINVESTAV and Its Relation to "the" French School

CINVESTAV's Department of Mathematics Education (DME) was created by Filloy and Imaz. Its founding reflected the urgent need to modernize study plans and programs for mathematics in harmony with an international movement which demanded that the mathematics being taught be brought more in line with current theories in the discipline. In that context, two French researchers, Brousseau and Pluvinage, were invited to Mexico City in 1979, thus initiating relations between Mexico and "the" French Didactic School.

Since its creation, the DME has maintained a well-defined scientific vocation. Because it is part of a centre devoted to research, its institutional conditions propitiate developing research projects, organizing and participating in congresses, colloquia and spaces for disseminating science, teacher-training courses, and Master's and Doctoral programs, among other scientific and academic activities. Also, it has a copious production of materials, both research-based and didactic (designed for teachers and students). Thus, it is an ideal space for establishing relations with research groups in other latitudes.

The first research projects developed at the DME centred on rational numbers, algebra, probability and functions, within theoretical frameworks that included epistemology, the cognitive sciences, and computation. Today, however, the fields explored have diversified broadly. Its methods for developing these activities and disseminating their results have been presented at international meetings (ICME, CIEAEM, PME) that led to forging contacts with research groups in France. The fact that two members of the DME—Hitt and Alarcón—did their doctoral studies in France, followed a few years later by the post-doctoral study periods of Cantoral and Farfán, have clearly influenced the development of Mathematics Education in Mexico through, for example, the creation of the *Escuela de Invierno en Matemática Educativa* (Winter School for Mathematics Education, EIME), reflecting the experience of French Summer School. A second key creation was the *Reunión Latinoamericana en Matemática Educativa* (Latin American Forum for Mathematics Education, RELME), which is held annually. The year 2017 has witnessed the 31st edition of this congress, which is opening up new channels for relations between France and Latin America, including a collection of studies framed in the Mathematical Working Space approach (see Sects. 2.1 and 2.2). Another event of this nature—propelled mainly by Brazilian researchers—is the *Simposio Latinoamericano de Didáctica de la Matemática* (Latin American Symposium on Mathematical Didactics, LADIMA[30]), which was held in November, 2016, with a strong presence of researchers from France and Latin America. Also, researchers at the DME were founding members of the *Comité Latinoamericano de Matemática Educativa* (Latin American Committee for Mathematics Education, CLAME), whose achievements include founding the *Revista Latinoamericana de Matemática Educativa* (Latin American Journal of Mathematics Education, RELIME), which appears in such indexes as ISI Thomson Reuters. Indeed, the journal's Editorial Board includes French researchers and its list of authors reflects joint studies conducted by French and Latin American scholars.

Finally, academic life at the DME is characterized by the enrolment of students from Mexico and several other Latin American countries in its Master's and Doctoral programs. These graduate students enjoy the opportunity to contact "the" French School through specialized literature, seminars, congresses and seminars with French researchers.

2.6.2 The DIE-CINVESTAV: A Strong Influence on Basic Education Supported by TDS

Since its founding, the Department of Educational Research (DIE for its initials in Spanish) has maintained a close relationship with Mexico's Department of Pub-

---

[30]http://www.boineventos.com.br/ladima, accessed January 2018.

lic Education that has allowed it to participate in producing textbooks, developing teaching manuals, and formulating study plans for math courses and processes of curriculum reform for basic education (students aged 6–15). The TDS has been the principal theoretical reference guiding these activities, as we show in the following section.

The 1980s brought the development of a project called "from six years" (*de los seis años*), one of the first TDS approaches, and one that would influence other programs, mainly in the curriculum reform for the area of basic education driven by the Department of Public Education in 1993. That reform introduced new textbooks, manuals and didactic activities strongly influenced by the French School, and especially TDS. Block, who would later complete his Doctorate (co-directed by Brousseau) participated actively in elaborating material for primary school, while Alarcón did the same for secondary school. That reform program was in place for 18 years (1993–2011), during a period that also saw the implementation of the Program for Actualizing Mathematics Professors (1995) and the introduction of several additional reforms that reflected the impact of TDS in Mexico's Teachers Colleges (*Escuelas Normales*), the institutions entrusted with training elementary school teachers. In Block's words, "after over 20 years of using those materials, it is likely that there are still traces of the contributions of TDS in the teaching culture".

2.6.3   The PROME at CICATA-IPN: A Professionalization Program for Teachers that Generates Relations Between France and Latin America

In the year 2000, the CICATA at Mexico's Instituto Politécnico Nacional created remote, online Master's and Doctorate Programs in Mathematics Education (known as PROME) designed for mathematics teachers. This program has generated multiple academic interactions between France and Latin America because, while students come mostly from Latin American nations (Mexico, Argentina, Chile, Colombia and Uruguay) some instructors are French. Also, their teacher training programs include elements of TDS and ATD, exemplified by the Study and Research Periods for Teacher Training (REI-FP).

The Learning Units (LU) on which these programs are based have been designed with the increasingly clear objective of functioning as a bridge between research in Mathematics Education and teaching practice. To this end, the organizers encourage a broad perspective on research and its results by including LUs designed by instructors at PROME and from other areas of the world. French researchers such as Athanaze (National Institute of Applied Sciences of Lyon, INSA-Lyon), Georget (Caen University), and Hache (LDAR and IREM-Paris), have participated in designing and implementing LUs, while Kuzniak (LDAR and IREM-Paris) was active in developing a Doctoral-level seminar, and Castela (LDAR and Rouen University) has participated in online seminars, workshops, the Doctorate's Colloquium, and PROME's inaugural Online Congress. These interactions allow the diffusion of research while recognizing teachers' professional knowledge and opening forums to answer urgent questions and identify demands that often develop into valuable research topics.

2.6.4   Theoretical Currents, Methodologies and Tools

Two theoretical currents that arose in Latin America—ethnomathematics and socioepistemology—have generated new ways of doing research and problematizing teaching and learning in the field of mathematics. The term ethnomathematics was coined by the Brazilian, D'Ambrosio, to label a perspective heavily influenced by Bishop's cultural perspective on mathematics education (Bishop, 1988). Socioepistemology, meanwhile, was first proposed by the Mexican scholar, Cantoral, and is now being developed by several Latin American researchers (see Cantoral, 2013). This approach that shares the importance attached to theoretical foundations in French research and the need for emancipation from the dominant traditions in this discipline, considers the epistemological role of social practices in the construction of mathematical knowledge; the mathematical object changes from being the focus of the didactical explanation to consider how it is used while certain normative practices take place. Since all kinds of knowledge matter—everyday and technical knowledge for example—Socioepistemology explains the permanent development of mathematical thinking considering not only the final mathematical production, but all those social circumstances—such as practices and uses—surrounding mathematical tasks.

Another theoretical development, though of narrower dimensions, is the extended praxeological model that emerged from Romo's doctoral dissertation (co-directed by Artigue and Castela), which has been widely disseminated in two publications (Romo-Vázquez, 2009; Castela & Romo, 2011). This approach makes it possible to analyse mathematical models in non-school contexts in order to transpose them to mathematics teaching by designing didactic activities shaped primarily for engineers and technicians; for example, the cases of electrical circuits and laminated materials (Siero & Romo-Vázquez, 2017), and blind source separation in engineering (Vázquez, Romo, Romo, & Trigueros, 2016). This model, accompanied by TDS tools, has also been applied in analyses of the practices of migrant child labourers in northern Mexico (Solares, 2012).

In terms of methodologies, it could be said that, as in the French School, research conducted in Mexico is largely qualitative in nature, though data-gathering and implicative analyses using the CHIC program developed by Gras are now being utilized. The methodology of didactical engineering (see Sect. 2.1) is still one of the most often employed, though rarely in its pure form. Socioepistemology is recognized by its use of Artigue (1990) and Farfán (1997).

2.6.5   *Areas of Opportunity and Perspectives*

The institutions introduced herein, and the brief historical profiles presented, reveal how interactions between France and Mexico have strengthened the development of Mathematics Education in Latin America. However, as Ávila (2016) points out, many challenges still need to be addressed, especially in terms of ensuring that all students have access to high-quality instruction in mathematics that will allow them to better understand and improve the world. This entails participating in specific types of education, including the following: in indigenous communities, with children of migrant workers, child labourers in cities, and children who lack access to technology, and in multi-grade schools, to mention only a few. In this regard, the research by Solares (2012) and Solares, Solares, and Padilla (2016) has shown how elements of TDS and ATD facilitate analysing and "valuing" the mathematical activity of child labourers in work contexts in Sonora and metropolitan Mexico City, and the design of didactic material that takes into account the knowledge and needs of these population sectors.

Finally, we consider that another important area of opportunity is generating didactic proposals for the Telesecundaria system (which serves mainly rural areas in Mexico) that, in the 2014 educational census escolar, represented 45.3% of all schools involved in teaching adolescents aged 12–15. This system requires a "tele-teacher" (classes are transmitted by television) and a teacher-monitor whose role is mainly to answer students' questions and resolve their doubts; though they were recently granted more autonomy and are now expected to develop didactic material on "transversal" topics for various subjects (e.g. physics, history, mathematics and Spanish). In this educational setting, many scholars feel that designing SRPs (see Sect. 2.1 on ATD) may be an optimal approach, since SRPs are characterized by their "co-disciplinary" nature. Designing and implementing such materials will make it possible to better regulate teaching practice and the formation of citizens who are capable of questioning the world. It is further argued that there are more general areas, such as the role of multimedia tools in students' autonomous work, large-scale evaluations, online math education, the study of cerebral activity associated with mathematical activities, and the nature of mathematical activities in technical and professional contexts, among other fields, that could lead to establishing new relations between France and Mexico to conduct research in these, so far, little-explored areas.

There is no doubt that both French and Mexican research have benefitted from this long term collaboration. Mexican research has enriched the perspectives of French didactics, especially with the development of socioepistemology and ethnomathe-matics, which have shown—beyond European logic—how autochthonous and native American communities produce mathematical knowledge, while highlighting the role of social practices in the social construction of knowledge. These approaches have produced studies in the ATD framework (Castela, 2009; Castela & Elguero, 2013). Likewise, studies of equality, and of education-at-a-distance, primarily for the professionalization of teachers, have developed very significantly in Latin America, and may bring about new relations that seek a greater impact on innovation in the teaching of mathematics in the classroom.

2.7 Didactic Interactions Between France and African Countries. The Case of Tunisia

The past decade has seen an important development of research in mathematics education across Francophone Africa, and the collaboration with French didacticians has played a decisive role in this development. Collaborations in mathematics education between Francophone African countries and France started in fact early after these countries got their independence, in the New Math period, with the support of the recently created IREM network and the INRP. Some IREMs were even created in African countries, for instance in Senegal as early as 1975 (Sokhna & Trouche, 2016). When doctorate programs in the didactics of mathematics opened in French universities in 1975 (see Sect. 2.1), African students entered these programs and prepared doctoral theses under the supervision of French researchers. The data collected for the preparation of the thematic afternoon at ICME-13 show that by 1985, eleven such theses had been already defended by students from four countries, and twenty more at the end of the 20th century, with nine countries represented. One can also observe the progressive development of co-supervision with African researchers, especially within the institutional system of co-tutoring doctorate. This is the case for sixteen of the twenty-five theses defended since 2000.

Research collaboration was also nurtured by the regular participation of African didacticians in the biannual Summer schools of didactics of mathematics (see Sect. 2.1) and, since the year 2000, by the tri-annual conferences of the Francophone Mathematical Space (EMF) created on the initiative of the French sub-commission of ICMI, the CFEM. One important aim of the creation of the EMF structure was indeed to favour the inclusion of Francophone researchers from non-affluent countries into the international community of mathematics education. EMF conferences alternate South and North locations, and three have already been held in Africa, in Tozeur (Tunisia) in 2003, Dakar (Senegal) in 2009, and Algier (Algeria) in 2015.

As evidenced by the four case studies regarding Benin, Mali, Senegal and Tunisia included in Artigue (2016), these collaborations enabled the creation of several master and doctorate programs in didactics of mathematics in Francophone African countries, and supported the emergence and progressive maturation of a community of didacticians of mathematics in the region. A clear sign of this maturation is the recent creation of ADiMA (Association of African didacticians of mathematics), the first conference of which was held at the ENS (Ecole Normale Supérieure) of Yaounde in Cameroon in August 2016; the second one is planned at the Institute of Mathematics and Physical Sciences of Dangbo in Benin in August 2018.

Tunisia is a perfect example of such fruitful interactions. In the next paragraphs, I (Faïza Chellougui) review them, from the first collaborations, in the seventies, involving the Tunisian association of mathematical sciences (ATSM), the French association of mathematics teachers (APMEP) and the IREM network until today. I show how these collaborations have nurtured the progressive maturation of a Tunisian community of didacticians of mathematics, today structured in the Tunisian association of didactics of mathematics (ATDM), and enriched the research perspectives in both countries. More details can be found in (Chellougui & Durand-Guerrier, 2016).

2.7.1   The Emergence of Didactic Interactions Between France and Tunisia

As just mentioned above, didactic interactions between Tunisia and France in mathematics stem from the relationships between the APMEP and the ATSM created in 1968, the oldest association of mathematics teachers in the Arab world and in Africa. The two associations indeed have a long term tradition of cross-invitation and participation in their respective "National days" meetings and seminars, of regular exchange of publications, etc. As early as 1977, Brousseau was invited to the national days of the ATSM. He presented the didactics of mathematics, its questions, concepts and research methods to a large audience of teachers, illustrating these with the research work he was developing on the teaching of rational and decimal numbers.

The IREM network then played an important role, especially the IREM of Lyon, thanks to its director, Tisseron. Tisseron had taught mathematics to pre-service mathematics teachers at the ENS in Tunis in the seventies, and also didactics at the ENS in Bizerte in the eighties when a didactic course was introduced in the preparation of secondary mathematics teachers. The decisive step for the development of didactics of mathematics as a research field in Tunisia occurred in fact in 1998, with the accreditation of a graduate program in the didactics of mathematics (DEA) at the Institute for higher education and continuous training in Tunis (ISEFC).[31] This accreditation had been prepared by Abdeljaouad, mathematician and historian of mathematics at the University of Tunis, who has played a crucial role in the emergence and development of didactic research in Tunisia, and Tisseron (see Abdeljaouad, 2009 for more details).

2.7.2   Development and Institutionalization

In order to set-up this program, a fruitful collaboration developed between the ISEFC and four research teams in French universities having doctorate programs in the didactics of mathematics: the LIRDHIST team (now S2HEP[32]) in the University

---

[31] http://www.isefc.rnu.tn/home.htm (accessed 2018/01/08).
[32] https://s2hep.univ-lyon1.fr (accessed 2018/01/08).

of Lyon 1 of which Tisseron was a member, the DIDIREM team (now LDAR[33]) in the University Paris 7, the Leibniz team in the University Grenoble 1, and the LACES[34] at the University of Bordeaux 2. Initially the DEA courses were taught by researchers from these teams, but gradually Tunisian scholars took them partially in charge. In 2006, the DEA turned into a Master program in the framework of the LMD (License-Master-Doctorate) reform of university. This institutional change led to modification of the organization of the program, and from the fall of 2010, ISEFC was empowered to offer a Master of research in didactics of science and pedagogy. The overall objective is ensuring a high level of training taking into account the multiple components of careers in science education. Eventually, a new Master in didactics of mathematics was set-up in October 2015, aiming to be innovative and open to the international community.

Another important step for the institutionalization of the field was the creation of the Tunisian association of didactics of mathematics (ATDM) in 2007. This association provides an institutional status to the young community of didacticians of mathematics and supports the dissemination of its research activities and results. For instance, it organizes an annual seminar to which contribute both well-known international and Tunisian researchers, to allow the regular diffusion of new or on-going research and to promote exchanges and debates within the didactic community.

Since the establishment of the DEA, fifteen doctoral theses and more than forty DEA or Master dissertations have been defended, most of these under the co-supervision of French and Tunisian researchers. Research reported in these theses and dissertations mainly addresses the teaching and learning of specific mathematics concepts, from elementary grades up to university. Important attention is paid to the epistemology and history of the concepts and domains at stake. The existence of Tunisian researchers specialized in the history of mathematics especially Arabic mathematics, such as Abdeljaouad, certainly contributes to nurture and instrument this attention. The main theoretical frameworks used are those of the French didactics, and especially ATD, TDS and TCF (see Sect. 2.1). One specificity however, is the number of theses and research projects that concern higher education and the transition from secondary to tertiary mathematics education (see for instance the theses by Chellougui on the use of quantifiers in university teaching (Chellougui, 2004), by Ghedamsi on the teaching of Analysis in the first university year (Ghedamsi, 2008), and by Najar on the secondary-tertiary transition in the area of functions (Najar, 2010). In fact, these theses and more global collaboration between French and Tunisian researchers have substantially contributed to the development of research in the area of higher education in France and Francophone countries in the last decade. This is also the case for logical perspectives in mathematics education, as shown by the thesis of Ben Kilani, Chellougui and Kouki (Chellougui, 2004; Ben Kilani, 2005; Kouki, 2008). Tunisian researchers have also pushed new lines of research, such as those related to the teaching and learning of mathematics in multilinguistic contexts, poorly addressed by French didacticians. In that area, a

---

[33] https://www.ldar.website (accessed 2018/01/08).

[34] http://www.laces.univ-bordeauxsegalen.fr (accessed 2018/01/08).

pioneering work was the thesis of Ben Kilani who used logic to show the differences of functioning of the negation in the Arabic and French language, and to understand the difficulties induced by these differences in the transition between Arabic and French as language of instruction in grade 9.

2.7.3 Some Outcomes of the French-Tunisian Didactic Collaboration

This long term collaboration has enabled the emergence, progressive development and institutionalization of didactics of mathematics as a field of research and practice in Tunisia. The majority of Ph.D. graduates have found a position in Tunisian higher education; they constitute today a community with the capacity of taking in charge the didactic preparation of primary and secondary mathematics teachers, and most Master courses. In 2017 moreover, the two first habilitations for research supervision[35] have been delivered to Tunisian didacticians (Ghedamsi, 2017; Kouki, 2017), which is a promising step for this community.

The French-Tunisian collaboration has certainly played a role in the increasing regional and international visibility and recognition of Tunisian researchers in the didactics of mathematics observed in the last decade. International visibility and recognition expresses through contributions to Francophone international events in the field such as the Summer school of didactics of mathematics to which the Tunisian delegation is regularly the largest foreign delegation, the EMF and ADiMa conferences. Recently it has also expressed through contributions to CERME conferences organized by the European Society for Research in Mathematics Education or the recently created INDRUM network, federating mathematics education research at university level. International recognition also expresses through invited lectures at seminars and congresses, and diverse scientific responsibilities. My personal case is a good illustration. I had an invited lecture at ICME-13 in 2016, and am a member of the International program committee of ICME-14. I have been a member of the scientific committee of two Summer schools, in 2007 and 2017, of EMF 2015 and INDRUM 2016. I was co-chair of the Topic Study Group entitled *"Pluralités culturelles et universalité des mathématiques: enjeux et perspectives pour leur enseignement et apprentissage"* (Cultural diversity and universality of mathematics: stakes and perspectives) at EMF2015, and of the group "Logic, numbers and algebra" at INDRUM 2016.

There is no doubt that there exists today a Tunisian community of didacticians of mathematics, dynamic and mature, open to the world beyond the sole frontiers of the Francophone world. While maintaining privileged links with the French didactic community, which has supported its emergence and development, it is creating its

---

[35] Habilitation for research supervision is a diploma compulsory to compete for full professor position at university in Tunisia as is the case in France.

own identity, and more and more offers challenging perspectives and contributions to this French community.

2.8 Epilogue

In this chapter reflecting the contributions at the ICME-13 thematic afternoon, we have tried to introduce the reader to the French didactic tradition, describing its emergence and historical development, highlighting some of its important characteristics, providing some examples of its achievements, and also paying particular attention to the ways this tradition has migrated outside the frontiers of the French hexagon and nurtured productive relationships with researchers in a diversity of countries, worldwide. This tradition has a long history, shaped as all histories by the conditions and constraints of the context where it has grown and matured. Seen from the outside, it may look especially homogeneous, leading to the term of French school of didactics often used to label it. The three theoretical pillars that have structured it from its origin and progressively developed with it, with their strong epistemological foundations, the permanent efforts of the community for maintaining unity and coherence, for capitalizing research knowledge, despite the divergent trends normally resulting from the development of the field, certainly contribute to this perception. We hope that this chapter shows that cultivating coherence and identity can go along with a vivid dynamics. The sources of this dynamics are to be found both in questions internal and external to the field itself and also, as we have tried to show, in the sources of inspiration and questions that French didacticians find in the rich connections and collaborations they have established and increasingly continue to establish both with close fields of research and with researchers living in other traditions and cultures. The first sections, for instance, have made clear the important role played by connections with psychology, cognitive ergonomics and computer sciences. The last four sections have illustrated the particular role played by foreign doctorate students, by the support offered to the creation of master and doctorate programs, and also by institutional structures such as the IREM network or the INRP, now IFé, since the early seventies; this is confirmed by the eight case studies presented in (Artigue, 2016). Looking more precisely at these international connections, there is no doubt that they are not equally distributed over the world. Beyond Europe, Francophone African countries, Vietnam, and Latin America are especially represented. For instance, among the 181 doctoral theses of foreign students supervised or co-supervised by French didacticians between 1979 and 2015, the distribution is the following: 56 from Francophone Africa, 54 from America, all but one from Latin America among with 28 for Brazil, 15 from East Asia among with 12 from Vietnam, 15 from Middle East among with 15 from Lebanon, and 36 from Europe.[36] For Francophone African countries, Vietnam and Lebanon, the educational links established

---

[36]Data collected by Patrick Gibel (ESPE d'Aquitaine) with the support of Jerôme Barberon (IREM de Paris).

in the colonial era are an evident reason. For Latin America, the long term cultural connections with France and the place given to the French language in secondary education until recently, and also the long term collaboration between French and Latin American mathematics communities, have certainly played a major role. These connections are dynamic ones, and regularly new ones emerge. For instance, due to the will of the French Ecoles normales supérieures (ENS) and to the East China Normal University (ECNU, Shanghai) to develop their collaboration in various domains of research (philosophy, biology, history, ... including education), and to the presence on each side of researchers in mathematics education, new links have emerged since 2015 and, currently, 4 Ph.D. are in preparation, co-supervised by researchers from the two institutions.

All these connections and collaborations allow us to see our tradition from the outside, to better identify its strengths and weaknesses, as pointed out by Knipping in Sect. 2.4, and to envisage ways to jointly progress, at a time when the need of research in mathematics education is more important than ever.

References

1. Abdeljaouad, M. (2009). L'introduction de la didactique des mathématiques en Tunisie. *Revue Africaine de didactique des sciences et des mathématiques, 4*, 1–14. http://www.radisma. infodocument.php?id=851.
2. Akkerman, S. F., & Bakker, A. (2011). Boundary crossing and boundary objects. *Review of Educational Research, 81*(2), 132–169.
3. Arcavi, A., Boero, P., Kilpatrick, J., Radford, L., Dreyfus, T., & Ruthven, K. (2016). Didactique goes travelling: Its actual and potential articulations with other traditions of research on the learning and teaching of mathematics. In B. R. Hodgson, A. Kuzniak, & J. B. Lagrange (Eds.), *The didactics of mathematics: Approaches and issues* (pp. 15–41). New York: Springer.
4. Arsac, G., & Durand-Guerrier, V. (2000). Logique et raisonnement mathématique. variabilité des exigences de rigueur dans les démonstrations mettant en jeu des énoncés existenciels. In T. Assude & B. Grugeon (Eds.), *Actes du séminaire national de didactique des mathématiques* (pp. 55–83). Paris: Equipe DIDIREM, Université Paris 7.
5. Artigue, M. (1990). Epistémologie et didactique. *Recherches en Didactique des Mathématiques, 10*(2/3), 241–286.
6. Artigue, M. (2002). Learning mathematics in a CAS environment: The genesis of a reflection about instrumentation and the dialectics between technical and conceptual work. *International Journal of Computers for Mathematics Learning, 7*, 245–274.
7. Artigue, M. (2014). Perspectives on design research: The case of didactical engineering. In A. Bikner-Ahsbahs, C. Knipping, & N. Presmeg (Eds.), *Approaches to qualitative research in mathematics education* (pp. 467–496). New York: Springer.
8. Artigue, M. (Ed.). (2016). *La tradition didactique française au-delà des frontières. Exemples de collaborations avec l'Afrique, l'Amérique latine et l'Asie.* Paris: CFEM. http://www.cfem.asso. fr/cfem/Collaborationsdidactiquesfrancaises.pdf. Accessed January 8, 2018.
9. Artigue, M., Assude, T., Grugeon, B., & Lenfant, A. (2001). Teaching and learning algebra: Approaching complexity through complementary perspectives. In H. Chick, K. Stacey, & J. Vincent (Eds.), *The Future of the Teaching and Learning of Algebra, Proceedings of the 12th ICMI Study Conference* (Vol. 1, pp. 21–32), Melbourne, Australia: The University of Melbourne.
10. Artigue, M., & Mariotti, M. A. (2014). Networking theoretical frames: The ReMath enterprise. *Educational Studies in Mathematics, 85*(3), 329–356.

11. Arzarello, F., Bazzini, L., & Chiappini, G. (2001). A model for analysing algebraic process of thinking. In R. Surtherland, T. Rojano, A. Bell, & R. Lins (Eds.), *Perspectives on school algebra* (pp. 61–81). New York: Springer.

12. Ávila, A. (2016). La investigación en educación matemática en México: una mirada a 40 años de trabajo. *Educación Matemática, 28*(3), 31–59.

13. Balacheff, N. (1987). Processus de preuve et situations de validation. *Educational Studies in Mathematics, 18,* 147–176.

14. Balacheff, N. (1988). *Une Étude des Processus de Preuve en Mathématique chez les Élèves de Collège.* Grenoble: Université Joseph Fourier.

15. Balacheff, N. (2013). cK¢, a model to reason on learners' conceptions. In M. V. Martinez, & A. Castro Superfine (Eds.), *Proceedings of PME-NA 2013—Psychology of Mathematics Education, North American Chapter* (pp. 2–15). Chicago, IL: University of Illinois at Chicago.

16. Balacheff, N., & Margolinas, C. (2005). cK¢ modèle de connaissances pour le calcul de situations didactiques. In A. Mercier & C. Margolinas (Eds.), *Balises en Didactique des Mathématiques* (pp. 75–106). Grenoble: La Pensée sauvage éditions.

17. Barbazo, E. (2010). *L'APMEP, un acteur politique, scientifique, pédagogique de l'enseignement secondaire mathématique du 20e siècle en France* (Doctoral thesis). Paris: EHESS.

18. Bartolini Bussi, M. G., & Mariotti, M. A. (2008). Semiotic mediations in the mathematics classroom: Artifacts and signs after a Vygotskian pertspective. In L. English, M. Bartolini Bussi, G. Jones, R. Lesh, & D. Tirosh (Eds.), *Handbook of international research in mathematics education, second revised version* (pp. 746–805). Mahwah: Lawrence Erlbaum.

19. Ben Kilani, I. (2005). *Les effets didactiques des différences de fonctionnement de la négation dans la langue arabe, la langue française et le langage mathématique* (Doctoral Thesis). Lyon: University of Tunis and University Claude Bernard Lyon 1.

20. Bishop, A. J. (1988). *Mathematical enculturation: A cultural perspective on mathematics education.* The Netherlands: Kluwer Academic Publishers.

21. Boero, P. (1994). Situations didactiques et problèmes d'apprentissage: convergences et divergences dans les perspectives de recherche. In M. Artigue, R. Gras, C. Laborde, & P. Tavignot (Eds.), *Vingt ans de Didactique des mathématiques en France* (pp. 17–50). Grenoble: La Pensée sauvage éditions.

22. Bolea, P., Bosch, M., & Gascón, J. (1999). The role of algebraization in the study of a mathematical organisation. In I. Schwank (Ed.), *European Research in Mathematics Education I: Proceedings of the First Conference of the European Society for Research in Mathematics Education* (Vol. II, pp. 135–145). Osnabrück: Forschungsinstitut fuer Mathematik didaktik.

23. Bosch, M. (2015). Doing research within the anthropological theory of the didactic: The case of school algebra. In S. J. Cho (Ed.), *Selected regular lectures from the 12th international congress on mathematical education* (pp. 51–69). New York: Springer.

24. Bosch, M., & Gascón, J. (2006). Twenty-five years of the didactic transposition. In B. Hodgson (Ed.), *ICMI bulletin no. 58* (pp. 51–65). https://www.mathunion.org/fileadmin/ICMI/files/Publications/ICMI_bulletin/58.pdf. Accessed January 8, 2018.

25. Brousseau, G. (1986). La relation didactique: le milieu. In *Actes de la IVème École d'été de didactique des mathématiques* (pp. 54–68). Paris: IREM Paris 7.

26. Brousseau, G. (1997). *Theory of didactical situations in mathematics.* Dordrecht: Kluwer Academic Publishers.

27. Brousseau, G., Brousseau, N., & Warfield, V. (2014). *Teaching fractions through situations: A fundamental experiment.* New York: Springer.

28. Bulf, C. (2008). *Étude des effets de la symétrie axiale sur la conceptualisation des isométries planes et sur la nature du travail géométrique au collège* (Doctoral Thesis). Université Paris Diderot—Paris 7. https://tel.archives-ouvertes.fr/tel-00369503. Accessed January 8, 2018.

29. Butlen, D., Bloch, I., Bosch, M., Chambris, C., Cirade, G., Clivaz, G., et al. (Eds.). (2015). *Rôles et places de la didactique et des didacticiens des mathématiques dans la société et dans le système éducatif.* Grenoble: La Pensée sauvage éditions.

30. Cantoral, R. (2013). *Teoría Socioepistemológica de la Matemática Educativa. Estudios sobre construcción social del conocimiento.* Barcelona: Gedisa.

31. Castela, C. (2009). La noción de praxeología: de la Teoría Antropológica de lo Didáctico (TAD) a la Socioepistemología. In P. Lestón (Ed.), *Acta Latinoamericana de Matemática Educativa* (Vol. 22, pp. 1195–1206). México, DF: Colegio Mexicano de Matemática Educativa A. C. y Comité Latinoamericano de Matemática Educativa A. C.

32. Castela, C., & Elguero, C. (2013). Praxéologie et institution, concepts clés pour l'anthropologie épistémologique et la socioépistémologie. *Recherches en Didactique des Mathématiques, 33*(2), 79–130.

33. Castela, C., & Romo, A. (2011). Des mathématiques à l'automatique: étude des effets de transposition sur la transformée de Laplace dans la formation des ingénieurs. *Recherches en Didactique des Mathématiques, 31*(1), 79–130.

34. Chaachoua, H. (2015). Étude comparative des recherches sur l'apprentissage de l'algèbre élémentaire: rapports croisés, bilan et perspectives. In D. Butlen, et al. (Eds.), *Rôles et places de la didactique et des didacticiens des mathématiques dans la société et dans le système éducatif* (pp. 21–39). Grenoble: La Pensée sauvage éditions.

35. Chellougui, F. (2004). *L'utilisation des quantificateurs universel et existentiel en première année universitaire entre l'explicite et l'implicite* (Doctoral thesis). Tunis, Lyon: University of Tunis and University Claude Bernard Lyon 1.

36. Chellougui, F., & Durand-Guerrier, V. (2016). Recherches en didactique des mathématiques en Tunisie. Collaborations avec la France. In M. Artigue (Ed.), *La tradition didactique française au-delà des frontières. Exemples de collaborations avec l'Afrique, l'Amérique latine et l'Asie*, (pp. 39–58). Paris: CFEM.

37. Chesnais, A. (2009). *L'enseignement de la symétrie axiale en sixième dans des contextes différents: les pratiques de deux enseignants et les activités des élèves* (Doctoral thesis). Paris: Université Paris Diderot—Paris 7. https://tel.archives-ouvertes.fr/tel-00450402. Accessed January 8, 2018.

38. Chesnais, A. (2012). L'enseignement de la symétrie orthogonale en sixième: des contraintes, des ressources et des choix. *Recherches en didactique des mathématiques, 32*(2), 229–278.

39. Chesnais, A., & Mathé, A.-C. (2015). Articulation between students' and teacher's activity during sessions about line symmetry. In K. Krainer, & N. Vondrová (Eds.), *Proceedings of the Ninth Congress of the European Society for Research in Mathematics Education* (pp. 522–528). Prague: Charles University in Prague.

40. Chevallard, Y. (1985a). *La transposition didactique. Du savoir savant au savoir enseigné.* Grenoble: La Pensée sauvage.

41. Chevallard, Y. (1985b). Le passage de l'arithmétique à l'algébrique dans l'enseignement des mathématiques au collège – première partie: l'évolution de la transposition didactique. *Petit x, 5*, 51–94.

42. Chevallard, Y. (1989). Le passage de l'arithmétique à l'algébrique dans l'enseignement des mathématiques au collège – deuxième partie: perspectives curriculaires: la notion de modélisation. *Petit x, 19*, 43–72.

43. Chevallard, Y. (1990). Le passage de l'arithmétique à l'algébrique dans l'enseignement des mathématiques au collège – Troisième partie. Perspectives curriculaires: voies d'attaque et problèmes didactiques. *Petit x, 23*, 5–38.

44. Chevallard, Y. (1994). Enseignement de l'algèbre et transposition didactique. *Rendiconti del seminario matematico Università e Politecnico Torino, 52*(2), 175–234.

45. Chevallard, Y. (1997). Familière et problématique, la figure du professeur. *Recherches en Didactique des Mathématiques, 17*(3), 17–54.

46. Chevallard, Y. (1999). L'analyse des pratiques enseignantes en théorie anthropologique du didactique. *Recherches en didactique des mathématiques, 19*(2), 221–266.

47. Chevallard, Y. (2015). Teaching mathematics in tomorrow's society: A case for an oncoming counter paradigm. In S. J. Cho (Ed.), *Proceedings of the 12th International Congress on Mathematical Education* (pp. 173–188). New York: Springer.

48. Chevallard, Y., & Bosch, M. (2014). Didactic transposition in mathematics education. In S. Lerman (Ed.), *Encyclopedia of mathematics education* (pp. 170–174). New York: Springer.

49. Chevallard, Y., & Sensevy, G. (2014). Anthropological approaches in mathematics education, French perspectives. In S. Lerman (Ed.), *Encyclopedia of mathematics education* (pp. 38–43). New York: Springer.

50. Choquet, G. (1964). *L'enseignement de la géométrie*. Paris: Hermann.

51. Coppé, S., & Grugeon-Allys, B. (2015). Étude multidimensionnelle de l'impact des travaux de recherche en didactique dans l'enseignement de l'algèbre élémentaire. Quelles évolutions? Quelles contraintes? Quelles perspectives? In D. Butlen et al. (Eds.), *Rôles et places de la didactique et des didacticiens des mathématiques dans la société et dans le système éducatif* (pp. 41–73). Grenoble: La Pensée sauvage éditions.

52. Coulange, L., Drouhard, J.-P., Dorier, J.-L., & Robert, A. (Eds.). (2012). *Enseignement de l'algèbre élémentaire: bilan et perspectives. Recherches en Didactique des Mathématiques (special issue)*. Grenoble: La Pensée sauvage éditions.

53. Cousin, M., & Miyakawa, T. (2017). Evolution of proof form in Japanese geometry textbooks. In T. Dooley, & G. Gueudet (Eds.) (2017, in preparation), *Proceedings of the Tenth Congress of the European Society for Research in Mathematics Education* (CERME10, February 1–5, 2017). Dublin, Ireland: DCU Institute of Education and ERME. https://keynote.conference-services. net/resources/444/5118/pdf/CERME10_0340.pdf. Accessed January 8, 2018.

54. Dieudonné, J. (1964). *Algèbre linéaire et géométrie élémentaire*. Paris: Hermann.

55. Douady, R. (1986). Jeux de cadres et dialectique outil-objet. *Recherches en Didactique des Mathématiques, 7*(2), 5–32.

56. Drijvers, P. (2013). Digital technology in mathematics education: Why it works (or doesn't). *PNA, 8*(1), 1–20.

57. Drijvers, P., & Trouche, L. (2008). From artifacts to instruments: A theoretical framework behind the orchestra metaphor. In K. Heid & G. Blume (Eds.), *Research on technology and the teaching and learning of mathematics* (Vol. 2, pp. 363–392), Cases and perspectives Charlotte, NC: Information Age.

58. Drouhard, J.-Ph. (2010). Epistemography and algebra. In V. Durand-Guerrier, S. Soury-Lavergne, & F. Arzarello (Eds.), *Proceedings of the 6th Congress of the European Society for Research in Mathematics Education* (pp. 479–486). Lyon: INRP.

59. Durand-Guerrier, V. (2013). Quelques apports de l'analyse logique du langage pour les recherches en didactique des mathématiques. In A. Bronner, et al. (Eds.), *Questions vives en didactique des mathématiques: problèmes de la profession d'enseignant, rôle du langage* (pp. 233–265). Grenoble: La Pensée sauvage éditions.

60. Duval, R. (1995). *Sémiosis et pensée humaine*. Bern: Peter Lang.

61. Farfán, R.-M. (1997). *Ingeniería Didáctica: Un estudio de la variación y el cambio*. Mexico: Grupo Editorial Iberoamérica.

62. Forest, D. (2012). Classroom video data and resources for teaching: Some thoughts on teachers' development. In G. Gueudet, B. Pepin, & L. Trouche (Eds.), *From text to "lived" resources. Mathematics curriculum materials and teacher development* (pp. 215–230). New York: Springer.

63. Ghedamsi, I. (2008). *Enseignement du début de l'analyse réelle à l'entrée à l'université: Articuler contrôles pragmatique et formel dans des situations à dimension a-dida*ctique (Doctoral thesis). Tunis, Bordeaux: University of Tunis and University Bordeaux 2.

64. Ghedamsi, I. (2017). *Modèles d'investigation des pratiques institutionnelles en mathématiques-Dispositifs méthodologiques en didactique des mathématiques, dans une perspective de conception d'ingénieries*. University Habilitation in Didactics of Mathematics. University of Tunis.

65. Gispert, H. (2014). Mathematics education in France, 1900–1980. In A. Karp & G. Schubring (Eds.), *Handbook on the history of mathematics education* (pp. 229–240). New York: Springer.

66. Gras, R. (1996). *L'implication statistique: nouvelle méthode exploratoire de données applications à la didactique* (avec la collaboration de Saddo Ag Almouloud, Marc Bailleul, Annie Larher, Maria Polo, Harrisson Ratsimba-Rajohn, André Totohasina). Grenoble: La Pensée sauvage éditions.

67. Grenier, D. (1988). *Construction et étude du fonctionnement d'un processus d'enseignement sur la symétrie orthogonale en sixième* (Doctoral thesis). Grenoble: Université Joseph Fourier, Grenoble 1.

68. Grugeon, B. (1995). *Etude des rapports institutionnels et des rapports personnels des élèves dans la transition entre deux cycles d'enseignement* (Doctoral thesis). Paris: Université Paris 7.
69. Gueudet, G., & Trouche, L. (2009). Towards new documentation systems for mathematics teachers? *Educational Studies in Mathematics, 71*(3), 199–218.
70. Gueudet, G., Pepin, B., & Trouche, L. (2013). Collective work with resources: An essential dimension for teacher documentation. *ZDM, The International Journal on Mathematics Education, 45*(7), 1003–1016.
71. Gueudet, G., Pepin, B., & Trouche, L. (Eds.). (2012). *From text to 'lived' resources: Mathematics curriculum materials and teacher development*. New York: Springer.
72. Guin, D., & Trouche, L. (1999). The complex process of converting tools into mathematical instruments. The case of calculators. *The International Journal of Computers for Mathematical Learning, 3*(3), 195–227.
73. Joffredo-Le Brun S., Morelatto, M., Sensevy, G. & Quilio, S. (2017). Cooperative engineering in a joint action paradigm. *European Educational Research Journal*. http://journals.sagepub.com/doi/pdf/10.1177/1474904117690006. Accessed January 8, 2018.
74. Knipping, C. (2003). *Beweisprozesse in der Unterrichtspraxis – Vergleichende Analysen von Mathematikunterricht in Deutschland und Frankreich*. Hildesheim: Franzbecker Verlag.
75. Kouki, R. (2008). *Enseignement et apprentissage des équations, inéquations et fonctions au Secondaire: entre syntaxe et sémantique* (Doctoral thesis). Lyon, Tunis: University Claude Bernard Lyon 1 and University of Tunis.
76. Kouki, R. (2017). *Recherches sur l'articulation des dimensions sémantiques, syntaxiques, sémiotiques, praxéologiques et épistémologiques dans l'enseignement et l'apprentissage des mathématiques. Etude de cas: algèbre du secondaire et développements limités au début de l'université*. University Habilitation in Didactics of Mathematics. University of Tunis.
77. Kuzniak, A., Tanguay, D., & Elia, I. (2016). Mathematical working spaces in schooling. *ZDM, 48*(6), 721–737.
78. Kynigos, C., & Lagrange, J.B. (Eds.) (2014). Special issue: Representing mathematics with digital media: Working across theoretical and contextual boundaries. *Educational Studies in Mathematics, 85*(3).
79. Laborde, C. (Ed.). (1988). *Actes du 1er colloque franco-allemand de didactique des mathématiques et de l'informatique*. Grenoble: La pensée sauvage éditions.
80. Laborde, C. (1995). Designing tasks for learning geometry in a computer based environment. In L. Burton & B. Jaworski (Eds.), *Technology in mathematics teaching-a bridge between teaching and learning* (pp. 35–68). London: Chartwell-Bratt.
81. Legrand, M. (2001). Scientific debate in mathematics courses. In D. Holton (Ed.), *The teaching and learning of mathematics at University Level*. New York: Springer.
82. Lagrange, J.-B., Artigue, M., Laborde, C., & Trouche, L. (2003). Technology and mathematics education: A multidimensional study of the evolution of research and innovation. In A. J. Bishop, M. A. Clements, C. Keitel, J. Kilpatrick, & F. K. S. Leung (Eds.), *Second international handbook of mathematics education* (pp. 239–271). Dordrecht: Kluwer Academic Publishers.
83. Lima, I. (2006). *De la modélisation de connaissances des élèves aux décisions didactiques des professeurs – Étude didactique dans le cas de la symétrie orthogonale* (Doctoral thesis). Université Joseph Fourier, Grenoble I. https://tel.archives-ouvertes.fr/tel-00208015. Accessed January 8, 2018.
84. Longo, G. (2012). Theorems as constructive visions. In G. Hanna & M. de Villiers (Eds.), *ICMI study 19 book: Proof and proving in mathematics education* (pp. 51–66). New York: Springer.
85. Margolinas, C. (2004). *Points de vue de l'élève et du professeur. Essai de développement de la Théorie des situations didactiques*. Note de synthèse (HDR). Marseille: Université de Provence. https://hal.archives-ouvertes.fr/tel-00429580. Accessed January 8, 2018.
86. Margolinas, C., Abboud-Blanchard, M., Bueno-Ravel, L., Douek, N., Fluckiger, A., Gibel, P., et al. (Eds.). (2011). *En amont et en aval des ingénieries didactiques. XVe école d'été de didactique des mathématiques*. Grenoble: La Pensée sauvage éditions.
87. Maschietto, M. (2008). Graphic calculators and micro-straightness: Analysis of a didactical engineering. *Int. J. Comput. Math. Learn., 13,* 207–230.

88. Maschietto, M. (2015). Teachers, students and resources in mathematics laboratory. In S. J. Cho (Ed.), *Selected Regular Lectures from the 12th International Congress on Mathematical Education* (pp. 527–546). New York: Springer.
89. Maschietto, M., & Soury-Lavergne, S. (2013). Designing a duo of material and digital artifacts: The pascaline and Cabri Elem e-books in primary school mathematics. *ZDM, 45*(7), 959–971.
90. Maschietto, M., & Trouche, L. (2010). Mathematics learning and tools from theoretical, historical and practical points of view: The productive notion of mathematics laboratories. *ZDM, 42*(1), 33–47.
91. Miyakawa, T. (2005). *Une étude du rapport entre connaissance et preuve: le cas de la notion de la symétrie orthogonale* (Doctoral thesis). Grenoble: Université Joseph Fourier, Grenoble I.
92. Najar, R. (2010). *Effets des choix institutionnels d'enseignement sur les possibilités d'apprentissage des étudiants. Cas des notions ensemblistes fonctionnelles dans la transition Secondaire/Supérieur* (Doctoral thesis). Tunis, Paris: Virtual University of Tunis and University Paris Diderot—Paris7.
93. Pedemonte, B. (2002). *Etude didactique et cognitive des rapports entre argumentation et démonstration dans l'apprentissage des mathématiques* (Doctoral thesis). Grenoble: Université Joseph Fourier, Grenoble.
94. Perrin-Glorian, M.-J. (1992). *Aires de surfaces planes et nombres décimaux. Questions didactiques liées aux élèves en difficulté aux niveaux CM-6ème* (Doctoral thesis). Paris: Université Paris 7.
95. Perrin-Glorian, M.-J. (2011). L'ingénierie didactique à l'interface de la recherche avec l'enseignement. Développement des ressources et la formation des enseignants. In C. Margolinas et al. (Eds.), *En amont et en aval des ingénieries didactiques* (pp. 57–74). Grenoble: La Pensée sauvage éditions.
96. Perrin-Glorian, M.-J., Mathé, A.-C., & Leclercq, R. (2013). Comment peut-on penser la continuité de l'enseignement de la géométrie de 6 à 15 ans? *Repères-IREM, 90,* 5–41.
97. Pilet, J., Chenevotot, F., Grugeon, B., El Kechaï, N., & Delozanne, E. (2013). Bridging diagnosis and learning of elementary algebra using technologies. In B. Ubuz, C. Haser, & M. A. Mariotti (Eds.), *Proceedings of the 8th Congress of the European Society for Research in Mathematics Education* (pp. 2684–2693). Ankara: Middle East Technical University and ERME.
98. Polo Capra, M. (1996). *Le repère cartésien dans les systèmes scolaires français et italien: étude didactique et application de méthodes d'analyse statistique multidimensionnelle.* Université Rennes 1.
99. Prediger, S., Bikner-Ahsbahs, A., & Arzarello, F. (2008). Networking strategies and methods for connecting theoretical approaches: First steps towards a conceptual framework. *ZDM, 40*(2), 165–178.
100. Rabardel, P. (1995/2002). *People and technology—A cognitive approach to contemporary instruments* https://hal.archives-ouvertes.fr/hal-01020705/document. Accessed January 8, 2018.
101. Robert, A., & Rogalski, J. (2002). Le système complexe et cohérent des pratiques des enseignants de mathématiques: une double approche. *Revue canadienne de l'enseignement des mathématiques, des sciences et des technologies, 2*(4), 505–528.
102. Robert, A., & Rogalski, J. (2005). A cross-analysis of the mathematics teacher's activity. An example in a French 10th-grade class. *Educational Studies in Mathematics, 59,* 269–298.
103. Romo-Vázquez, A. (2009). *La formation mathématique des futurs ingénieurs* (Doctoral thesis). Université Paris-Diderot. https://tel.archives-ouvertes.fr/tel-00470285/document. Accessed January 8, 2018.
104. Ruiz-Munzón, N., Bosch, M., & Gascón, J. (2007). The functional algebraic modelling at Secondary level. In D. Pitta-Panzati & G. Philippou (Eds.), *Proceedings of the 5th Congress of the European Society for Research in Mathematics Education* (pp. 2170–2179). Nicosia: University of Cyprus.
105. Ruiz-Munzón, N., Bosch, M., & Gascón, J. (2013). Comparing approaches through a reference epistemological model: The case of school algebra. In B. Ubuz, C. Haser, & M. A. Mariotti (Eds.), *Proceedings of the 8th Congress of the European Society for Research in Mathematics Education* (pp. 2870–2879). Ankara: Middle East Technical University and ERME.

106. Ruiz-Munzón, N., Matheron, Y., Bosch, M., & Gascón, J. (2012). Autour de l'algèbre: les entiers relatifs et la modélisation algébrico-fonctionnelle. In L. Coulange, J.-P. Drouhard, J.-L. Dorier, & A. Robert (Eds.), *Enseignement de l'algèbre élémentaire. Bilan et perspectives. Recherches en Didactique des Mathématiques, special issue* (pp. 81–101). Grenoble: La Pensée sauvage éditions.
107. Sekiguchi, Y., & Miyazaki, M. (2000). Argumentation and mathematical proof in Japan. La Lettre de la Preuve. http://www.lettredelapreuve.org/OldPreuve/Newsletter/000102Theme/000102ThemeUK.html. Accessed January 8, 2018.
108. Siero, L., & Romo-Vázquez, A. (2017). Didactic sequences teaching mathematics for engineers with focus on differential equations. In M. S. Ramírez & M. A. Domínguez (Eds.), *Driving STEM learning with educational technologies* (pp. 129–151). USA: IGI Global.
109. Sokhna, M., & Trouche, L. (2016). Repenser la formation des enseignants en France et au Sénégal: une source d'interactions fécondes. In M. Artigue (Ed.), *La tradition didactique française au-delà des frontières. Exemples de collaborations avec l'Afrique, l'Amérique latine et l'Asie* (pp. 27–38). Paris: CFEM.
110. Solares, D. (2012). *Conocimientos matemáticos de niños y niñas jornaleros agrícolas migrantes* (Doctoral thesis). Departamento de investigaciones Educativas Cinvestav-IPN, México.
111. Solares, D., Solares, A. & Padilla, E. (2016). La enseñanza de las matemáticas más allá de los salones de clase. Análisis de actividades laborales urbanas y rurales. *Educación Matemática, 28*(1), 69–98.
112. Spagnolo, F. (1995). *Les obstacles épistémologiques: le postulat d'Eudoxe-Archimède* (Doctoral thesis). Université Bordeaux I.
113. Tahri, S. (1993). *Modélisation de l'interaction didactique: un tuteur hybride sur CABRIGÉOMÈTRE pour l'analyse de décisions didactiques* (Doctoral thesis). Grenoble: Université Joseph Fourier, Grenoble.
114. Tavignot, P. (1993). Analyse du processus de transposition didactique. Application à la symétrie orthogonale en sixième lors de la réforme de 1985. *Recherches en didactique des mathématiques, 13*(3), 257–294.
115. Trgalová, J., & Chaachoua, H. (2009). Relationship between design and usage of educational software: The case of Aplusix. In V. Durand-Guerrier, S. Soury-Lavergne, & F. Arzarello (Eds.), *Proceedings of the 6th Congress of the European Society for Research in Mathematics Education* (pp. 1429–1438). Lyon: INRP.
116. Trouche, L. (2016). Didactics of mathematics: Concepts, roots, interactions and dynamics from France. In J. Monaghan, L. Trouche, & J. M. Borwein (Eds.), *Tools and mathematics, instruments for learning* (pp. 219–256). New York: Springer.
117. Vandebrouck, F. (Ed.). (2013). *Mathematics classrooms: Students' activities and teachers' practices*. Rotterdam: Sense Publishers.
118. Vázquez, R., Romo, A., Romo, R., & Trigueros, M. (2016). Una reflexión sobre el rol de las matemáticas en la formación de ingenieros. *Educación Matemática, 28*(2), 31–57.
119. Vergnaud, G. (1981). Quelques orientations théoriques et méthodologiques des recherches françaises en didactique des mathématiques. *Recherches en didactique des mathématiques, 2*(2), 215–232.
120. Vergnaud, G. (1991). La théorie des champs conceptuels. *Recherches en didactique des mathématiques, 10*(2/3), 133–170.
121. Vergnaud, G. (2009). The theory of conceptual fields. *Human Development, 52,* 83–94.

Chapter 3
Didactics of Mathematics in the Netherlands

Marja Van den Heuvel-Panhuizen

Abstract This chapter highlights key aspects of the didactics of mathematics in the Netherlands. It is based on the Dutch contribution to the Thematic Afternoon session on European didactic traditions in mathematics, organised at ICME13 in Hamburg 2016. The chapter starts with a section in which mathematics education in the Netherlands is viewed from four perspectives in which subsequently attention is paid to the role of mathematics and mathematicians, the role of theory, the role of design, and the role of empirical research. In all these themes Hans Freudenthal has played a key role. Hereafter, the focus is on two Dutch mathematics educators (Adri Treffers for primary school and Jan de Lange for secondary school) who each left an important mark on how the didactics of mathematics has developed in the last half century and became known as Realistic Mathematics Education (RME). To illustrate the principles of this domain-specific instruction theory a concrete task is worked out in the section "Travelling to Hamburg". The chapter concludes with five sections featuring voices from abroad in which mathematics educators from other countries give a short reflection on their experiences with RME.

Keywords Realistic mathematics education · IOWO · Freudenthal · Mathematisation · Mathematics as a human activity · Didactical phenomenology · Empirical didactical research · Treffers · Students' own productions and constructions · De Lange · Contexts for introducing and developing concepts · Design-based research · Task design · Parametric curve · Findings from field tests

M. Van den Heuvel-Panhuizen (✉)
Utrecht University, Utrecht, The Netherlands
e-mail: m.vandenheuvel-panhuizen@uu.nl; m.vandenheuvel-panhuizen@nord.no

M. Van den Heuvel-Panhuizen
Nord University, Bodø, Norway

3.1 Mathematics Education in the Netherlands Viewed from Four Perspectives

Marja van den Heuvel-Panhuizen

3.1.1 The Role of Mathematics and Mathematicians in Mathematics Education in the Netherlands

When mathematics became a compulsory subject in primary and secondary school in the Netherlands at the start of the 19th century, professional mathematicians were not much involved. Furthermore, the Dutch government took a somewhat restrained position on what mathematics was taught and particularly on how it was taught. Decisions regarding the curriculum were considered an internal school affair. In 1917 this policy was formalized in the Dutch Constitution as Freedom of Education. In practice, this meant that changes in the school mathematics curriculum were usually discussed by teacher unions, or special committees of teachers, then approved by the school inspectors and only after that ratified by the government. Professional mathematicians played hardly any formal part in this process. They sometimes participated in secondary education final examinations, by acting as assessors in the oral examinations and checking the grading of students' work on the written examinations, but they were not responsible for these examinations, which were devised by a selected group of teachers and approved by school inspectors. Furthermore, professional mathematicians were scarcely involved in the production of textbooks for secondary and primary mathematics education, which were mainly written by teachers. The government left the production of textbooks to the market. Schools were free to choose those books they liked most. Although there is currently more government involvement in the 'what' of teaching through formulating standards and a series of compulsory tests, the freedom regarding textbooks still exists.

Professional mathematicians' interest in school mathematics began to grow in the first half of the 20th century. For example, in 1924 a mathematics journal was extended by an addendum in which didactical questions could be discussed; this addendum later became the still existing journal *Euclides*. The first issue of this journal paid attention to the argument between the mathematician Dijksterhuis and the physicist Ehrenfest-Afanassjewa. The latter had also studied mathematics in Göttingen with Klein and Hilbert. While Dijksterhuis insisted on keeping to the traditional approach to teaching school geometry, which was based on the formal characteristics of the discipline, Ehrenfest argued for making use of students' intuitive knowledge and starting with concrete activities in three-dimensional space. However, Ehrenfest's ideas and those of her discussion group on teaching mathematics which was to become the Mathematics Working Group in 1936, did not find much response from mathematicians. Even the famous Dutch mathematician Brouwer did not have any affinity with teaching mathematics in school.

Yet, from 1950 onwards, more attention was paid to mathematics education within the community of mathematicians. This was reflected, for example, in the establishment in 1954 of the Dutch Education Commission for Mathematics, an ICMI subgroup, in which mathematics teachers and mathematicians cooperated. Freudenthal became chair of the group shortly after its foundation. Although in those days more mathematicians felt that mathematics teaching needed modernisation, it was Freudenthal in particular who had a genuine and deep interest in the didactics of mathematics. So, it is no wonder that, after World War II, he had also joined Ehrenfest's group.

In 1961, the Dutch government appointed the Commission Modernisation of the Mathematics Curriculum (CMLW), which was a new phenomenon in the long history of government that was rather aloof with respect to curriculum issues. The founding of this new commission, consisting of professional mathematicians and teachers, was a direct consequence of the Royaumont conference. The Dutch government became convinced of the urgency of the modernization of mathematics education. Freudenthal was the most outstanding mathematician in this commission and he was also the only one who was heavily involved in the didactics of mathematics. He convinced the other commission members to focus on the teaching of mathematics instead of on the content of the curriculum and he also moved the attention of the commission to the lower grades of schooling. Later on, in 1971, IOWO (Institute for Development of Mathematics Education) was established with Freudenthal as its first director. The opening of the institute was the beginning of a long period of over forty years in which mathematics teachers and educators worked on the design and research of mathematics education in primary and secondary school and teacher education. The instructional designs and the underlying theory of Realistic Mathematics Education (RME) developed at this institute have changed the Dutch mathematics curriculum and the approach to teaching mathematics, and this happened without any government interference. Characteristic of this approach is that it starts with offering students problems in meaningful situations, from which contexts can gradually evolve into models that can be used to solve a broader scope of problems; through the process of progressive schematisation, students eventually end up understanding mathematics at a more formal level.

Until the late 1990s, RME was generally accepted for primary and secondary education. Also university mathematicians were involved in secondary education reform projects such as HEWET and PROFI. However, after 2000 some university mathematicians started to blame RME for the lack of basic mathematical skills of their first-year students. They wanted to return to the way of teaching mathematics that (in their view) was common some forty years ago. For primary school mathematics education, the Ministry of Education and the Netherlands Royal Academy of Sciences (KNAW) appointed a commission to arbitrate this Dutch Math War. The commission's conclusion was that there was no evidence that students' achievements would be better with either RME or the mechanistic back-to-basics approach. This conclusion resulted in RME being less in the firing line, and it became possible again to have a professional discussion among all stakeholders about primary school mathematics education. In the area of secondary education, the Ministry of Education appointed the Commission Future Mathematics Education (cTWO), consisting of mathematicians, mathematics education researchers and mathematics teachers, to revise the mathematics curriculum for upper secondary mathematics education. In addition, Regional Support Centres (Bètasteunpunten) were established to facilitate connections between secondary schools and universities. Also, the Mathematical Society and the mathematics teachers' association (NVvW), set up Platform Mathematics Netherlands, a new organisation for collaboration, which included, among other things, a commission for mathematics education. This commission only covers secondary mathematics education and not the teaching of mathematics in primary school.

Background information about the role of mathematics and mathematicians in mathematics education in the Netherlands can be found in Goffree, Van Hoorn, and Zwaneveld (2000), La Bastide-van Gemert (2015) and Smid (2018), for example.

3.1.2   The Role of Theory in Mathematics Education in the Netherlands

Mathematics education is not just about the process of teaching mathematics, but also encompasses ideas and knowledge about how students learn mathematics, about how mathematics can best be taught and what mathematical content should be taught and why. Finding answers to these questions is the main goal of the scientific discipline that in the Netherlands—in line with the European tradition—is called the didactics of mathematics.

At the beginning of the 19th century when the first textbooks were published in the Netherlands, the prefaces of these textbooks showed the initial efforts towards a theory of mathematics education that contributed to the development of the didactics of mathematics as a scientific discipline. A next step forward came in 1874 when the Dutch schoolteacher Versluys published his book on methods for teaching mathematics and for the scientific treatment of the subject. However, a decisive move towards a theoretical basis of mathematics education in the Netherlands was Freudenthal's unfinished manuscript *Rekendidactiek* (Arithmetic Didactics), written in 1944, but never published. Freudenthal's interest in mathematics education in primary school was triggered during World War II when he was teaching arithmetic to his sons, and observed their learning processes. He also carried out an extensive literature review of the didactics of arithmetic. A further advancement in Freudenthal's thinking about mathematics education occurred at the end of the 1950s when he worked with the Van Hieles and became familiar with the theory of levels. Inspired by this, he developed the very important didactic principle of guided reinvention in which decisions about guidance should be informed by analysing learning processes. For Freudenthal, the 're' in reinvention points to students' learning processes, and the adjective 'guided' to the instructional environment. Viewing learning as guided reinvention means striking a subtle balance between the students' freedom of invention and the power of the teachers' guidance.

Freudenthal's intention of giving mathematics education a scientific basis resulted in the publication of *Weeding and Sowing* in 1978, which he called a preface to a science of mathematics education. In this book, he introduced the didactical phenomenological analysis of mathematics, an approach which was further elaborated in *Didactical Phenomenology of Mathematical Structures*, published in 1983. According to Freudenthal, thorough analysis of mathematical topics is needed in order to show where the student might step into the learning process of mankind. In other words, a didactical phenomenology, rather than a pure epistemology of what constitutes mathematics, is considered to inform us on how to teach mathematics. This phenomenology includes how mathematical 'thought objects' can help organising and structuring phenomena in reality, which phenomena may contribute to the development of particular mathematical concepts, how students can come into contact with these phenomena, how these phenomena beg to be organised by the mathematics intended to be taught, and how students can be brought to higher levels of understanding.

Although mathematics plays a central role in these analyses, Freudenthal rejected the idea of taking the structure of mathematics or the expert knowledge of mathematicians as his point of departure. The goal was making mathematics accessible and understandable for students by taking their learning processes seriously. Freudenthal viewed working on the design of education and experience with educational practice as necessary requirements for making theory development possible. His work at IOWO, which was founded in 1971, and particularly his collaboration with the Wiskobas group around Treffers, and later with the Wiskivon group for secondary education, which both did a great deal of work with students and teachers in schools, was therefore crucial for Freudenthal's thinking.

At the same time, however, Freudenthal's involvement was important for IOWO as well. In addition to promoting mathematisation and mathematics as a human activity that is connected to daily life or an imagined reality, and emphasizing that students and even young children can generate a large amount of mathematical thinking—yet in an informal context-connected way—Freudenthal's participation was essential for another reason too. Being an authority in the field of mathematics as a discipline, Freudenthal legitimised the work done at IOWO from the perspective of mathematics.

Although the activities at IOWO, due to its focus on designing education, could be characterised as engineering work rather than as research, IOWO produced 'valuable splinters' which could be counted as research output. Freudenthal saw this approach as paradigmatic for how theory development must take place: from designing educational practice to theory. The theory that evolved from this work at IOWO was later called Realistic Mathematics Education (RME) and was initially described by Treffers in 1978, and published in 1987 in his book *Three Dimensions*. The principles of this domain-specific instruction theory have been reformulated over the years, including by Treffers himself, but are presently still seen as leading for RME.

The first principle, the activity principle, follows from the interpretation of mathematics as a human activity, and implies that students are treated as active participants in the learning process. The reality principle arises from considering mathematics as based in reality and developing from it through horizontal mathematisation, and entails that mathematics teaching provides students with meaningful problems that can be mathematised. The level principle highlights the idea that students pass through several stages of understanding, from informal, context-connected to formal mathematics. In this process, didactical models serve a bridging function and vertical mathematisation is stimulated. This level principle is also reflected in the procedure of progressive schematization. The intertwinement principle states that the mathematics curriculum is not split into isolated strands, but that, following the mathematisation of reality, the focus is on the connection and coherence of mathematical structures and concepts. The interactivity principle signifies the social-cognitive aspect of learning mathematics, and entails that students are offered opportunities to share their thinking with others in order to develop ideas for improving their strategies and to deepen their understanding through reflection. The guidance principle refers to organising education in such a way that guided reinvention is possible, through a proactive role of the teacher and educational programs based on coherent long-term teaching-learning trajectories that contain scenarios which have the potential to work as a lever to effect shifts in students' understanding. Several local instruction theories focusing on specific mathematical topics have been developed which align with these general principles of RME.

RME is not a fixed and finished theory of mathematics education and is still in development. Over the years the successors of IOWO—OW&OC (Research of Mathematics Education & Education Computer Centre), the Freudenthal Institute and the lately established Freudenthal Group[1]—have made different emphases. As a result of collaboration with researchers in other countries, RME has also been influenced by theories from abroad such as social constructivist approaches which contributed to the interaction principle of RME and provided RME with a lens for investigating classroom discourse. More recently, elicited by the use of new technology in mathematics education, approaches inspired by instrumentation theory have connected with RME to achieve a better understanding of how tool use and concept development are related. Finally, a further new avenue is the revitalisation of the activity principle of RME through the incorporation of embodied cognition and perception-action theories in which the focus is also on how students' concept development and deep learning can be understood and fostered.

However, when working on the further development of RME through integrating it with other theories, it is still important that mathematics should maintain its central place. Developing mathematics education and investigating learning and teaching processes should always be grounded in mathe-didactical analyses which unpack mathematics in didactic terms and take into account phenomenological, genetic-epistemological, and historical-cultural perspectives.

---

[1]In 2012 the Freudenthal Institute was split into two. The research and design work in early childhood, primary education, special education, and intermediate vocational education were moved to the Faculty of Social and Behavioural Sciences (FSW) and carried out by the *Freudenthal Group*. The research and design work in secondary education have remained part of the *Freudenthal Institute* in the Faculty of Science.

Background information about the role of theory in mathematics education in the Netherlands can, for example, be found in De Lange (1987), Freudenthal (1991), Treffers (1987a), and Van den Heuvel-Panhuizen and Drijvers (2014).

3.1.3   The Role of Design in Mathematics Education in the Netherlands

Making things work, looking for pragmatic solutions and being creative and innovative are typical features of Dutch culture and they occupy an important place in Dutch society. This emphasis on design can also be recognized in mathematics education and can be considered the most significant characteristic of the Dutch didactic tradition in the past half century.

The reform movement in mathematics education that started in the Netherlands at the end of the 1960s was all about designing 'new' education, which in those days meant working on an alternative to the mechanistic approach to teaching mathematics that was prevalent in the Netherlands at the time. This approach, which still has some followers today, is characterised by teaching mathematics at a formal level from the outset, an atomized and step-by-step way of teaching in which the teacher demonstrates how problems must to be solved, and the scant attention paid to the application of mathematics. At the same time that the need arose for an alternative for this mechanistic approach, two new approaches from abroad appeared: the empiricist trend in which students were set free to discover a great deal by themselves and were stimulated to carry out investigations, and the structuralistic trend propagated by the New Math movement in which the mathematics to be taught was directly derived from mathematics as a discipline. However, neither of these new approaches was well received in the Netherlands.

Therefore, at the end of the 1960s the Wiskobas group started to think about another way to improve teaching mathematics in primary school. From 1971 on this took place at the newly-established IOWO, which some time later was extended to include the Wiskivon group that had been formed to design a new approach to teaching mathematics in secondary education. All staff members of these two groups, except Freudenthal, had experience in school practice either as a mathematics teacher or as a mathematics teacher educator. This meant that their work was very practice-oriented. The theory development which resulted in RME was considered a derivative of this practical work and would later serve as a guide for further design activities. Because of the focus on the practice of teaching, it is not surprising that Freudenthal often stated that IOWO was not a research institute, and IOWO staff members did not regard themselves as researchers, but as producers of instruction, as engineers in the educational field. As implied by the latter term, this work was not done in isolation, but carried out with teachers and students in classrooms. Moreover, there was a strong collaboration with mathematics teacher educators, counsellors at teacher advisory centres, and textbook authors with whom the materials were discussed and who also contributed to their development. In this way, the implementation of the reform happened more or less naturally without specific government interference. By having these strong networks of people and institutions involved in mathematics education, new ideas for teaching mathematics could immediately be used in pre-service teacher education, in-service courses, and above all in textbooks. Of all the possible change agents, textbooks have played a key role in the reform of mathematics education in the Netherlands. For primary school mathematics, the same is also true for the mathematics education infrastructure that evolved from these networks. For secondary mathematics education a teachers' association (NVvW) had already been founded in 1925 and for primary school mathematics the infrastructure came into being later. In 1981 Panama was set up, which has come to involve a collaboration of institutions for pre-service and in-service mathematics teacher education, and in 1982 NVORWO was established as an association for primary school mathematics. The main purpose of the infrastructure as a whole was, and still is, to inform the mathematics education community in the Netherlands through national mathematics education conferences, professional journals, in-service courses and websites and to support national mathematical events for students.

The educational designs that have been produced over the years by IOWO and its respective successors, are multifaceted, ranging from tasks containing opportunities for mathematisation and paradigmatic contexts that evolve into level-shifting didactical models, to tasks for mathematics days and competitions for students, to elaborate teaching sequences for particular mathematical domains. Among other things, the design work in primary school mathematics led to helpful contexts such as the pizza context in which students could produce fractions by themselves through fair sharing activities, and the bus context in which students were encouraged to reason about passengers entering and exiting and so invent their own symbolic notations of what happens at a bus stop. The design work for primary school also resulted in some very powerful didactical models that can be found in most current textbooks in the Netherlands, such as the empty number line, the arithmetic rack, the percentage bar and the ratio table. With respect to upper secondary education, new programs were developed in the 1980s and 1990s for Mathematics A (preparing students for studies in the social sciences) and Mathematics B (preparing students for studies in the natural sciences). Additionally, at the turn of the century new RME-based modules on calculus and geometry were developed for Mathematics B in the upper grades of pre-university secondary education. A prominent design project that was carried out with the University of Wisconsin involved the development of a complete textbook series *Mathematics in Context* for Grade 5–8 of the U.S. middle school. This project began in the mid-1990s and ran for some ten years.

Another long-term design project was the TAL project that started in 1997. Its aim was the development of longitudinal conceptual teaching-learning trajectories that describe the pathway that students largely follow in mathematics from Kindergarten to Grade 6. The decision to work on such trajectories was innovative at that time. The basis for this TAL project was the so-called Proeve, a first version of a national curriculum for primary school mathematics that led to the official enactment of the first description of the core goals for mathematics at the end of primary school at the beginning of the 1990s. The teaching-learning trajectories were designed to describe how these core goals could be reached, thus providing teachers, textbook authors and test developers with an insight into the continuous learning line of learning mathematics, so contributing to making the curriculum more coherent.

The advent of computer-based technology in schools again brought new demands and challenges for design. In addition to exploring opportunities for computer-assisted instruction, much effort was also put into rethinking the subject of mathematics within the context of the virtual world and exploring how students could benefit from the dynamic and interactive qualities of the new technology. This led not only to the development of the so-called Digital Mathematics Environment in which teachers can adapt and design instructional material for their students including the use of mathematical tools and feedback, but also resulted in a seemingly inexhaustible flow of applets and mini-games for primary and secondary education that are freely available online.

Background information about the role of design in mathematics education in the Netherlands can, for example, be found in Bakker (2004), Doorman (2005), Drijvers (2003), Gravemeijer (1994), National Center for Research in Mathematical Sciences Education & Freudenthal Institute (1997–1998), Streefland (1993), and Van den Heuvel-Panhuizen (1996, 2003).

3.1.4   The Role of Empirical Research in Mathematics Education in the Netherlands

Research in the Netherlands into the learning and teaching of mathematics since the first half of the 20th century has always been empirical in one way or another. Initially this research was undertaken mostly by psychologists and pedagogues with an interest in mathematics, but later on it was also done by mathematics teachers. A prominent example of such research was the didactical experiment carried out by Van Hiele-Geldof in the 1950s on teaching geometry in the first year of secondary school. Her thesis about this experiment contained a very careful description of how she developed the teaching sequence that brought students from visually supported thinking to abstract thinking. She also recorded precise protocols of what happened in the classroom, which were then thoroughly analysed. Starting with what she called a psychological-didactical analysis of the mathematical content was part of her research method. In fact, Van Hiele-Geldof's work, greatly admired by Freudenthal, contained many important ingredients of the research into the learning and teaching of mathematics that was done in the Netherlands from then on.

Freudenthal's empirical didactical research began with observing his own children as he was teaching them arithmetic during World War II. It is noteworthy that he warned at first against overestimating the value of these observations, stating that he would like to do observations with a more diverse sample and on a larger scale. Later, he apparently changed his opinion. In the 1970s he emphasized the strengths of qualitative small case studies. He even called on research in the natural sciences ("One Foucault pendulum sufficed to prove the rotation of the earth") to prove the power of observing a student's learning process. For Freudenthal, one good observation was worth more than hundreds of tests or interviews. The reason for this preference was that observing learning processes led to the discovery of discontinuities in learning, which Freudenthal regarded as being of great significance for understanding how students learn mathematics. Merely comparing scores of a large sample of students collected at different measuring points would imply that only an average learning process, in which the discontinuities have been extinguished and all essential details have disappeared, can be analysed.

The design work that was carried out at the IOWO from 1971 onwards, with the aim of creating materials and teaching methods for the reform of Dutch mathematics education, was also highly informed by empirical research. In agreement with Freudenthal, the emphasis was on small-scale qualitative studies carried out in schools. Based on didactical-phenomenological analyses of the mathematical domains and making use of knowledge of students' learning processes and the classroom context, learning situations were initially designed using thought experiments. These were followed by actual experiments with students and teachers, and the reactions of both students and teachers were observed. In this way, IOWO staff members ran school experiments in which the mathematics to be taught and the teaching methods were continuously adapted based on experiences in classrooms and feedback and input from the teachers. In this development process, design, try-out, evaluation and adaptation followed each other in short, quick cycles. Reflections on what happened in the classrooms focused not only on whether or not a learning process had taken place, but also on what impeded or facilitated its occurrence. These reflections and the accompanying intensive deliberations among IOWO staff members about the designed learning situations provided important theoretical insights which evolved into the RME theory of mathematics education. In their turn, these theoretical insights led further designs. In other words, theory development and the development of education were strongly interwoven.

This type of research, initially called developmental research but later given the internationally more common name design research, was the backbone of RME-based research activities. Over the years, the method of design research was developed further. Data collection and analysis procedures which would contribute to the evidential value of the findings of design research were added. The theoretical grounding was also elaborated, through prior mathe-didactical analyses and through including findings and approaches from the education and learning sciences.

In addition to design research, depending on what specific research questions have to be answered, various other empirical research methods are used, such as quasi-experiments (including pretest-posttest intervention designs and micro-genetic designs), surveys (including questioning teachers about their classroom practice and beliefs about mathematics education and carrying out expert consultations), document studies (including textbook and software analyses and study on the history of mathematics education) and review studies and meta-analyses. Also, outside the circle of RME-affiliated didacticians, there is a large group of researchers in the Netherlands consisting of psychologists, orthopedagogues, and cognitive neuroscientists who focus particularly on investigating how specific student characteristics influence students' learning of mathematics. In this way, they complement the research done by didacticians. Similarly, research by educationalists who, among other things, investigate school organisation and classroom climate also provide relevant knowledge for all involved in mathematics education.

Relevant empirical data to direct the development of mathematics education were also acquired through the PPON studies that, from 1987 on, have been carried out every five years by Cito, the national institute for educational measurement. It is important that these studies gave an overview of changes over time in the mathematics achievements of Dutch primary school students and of the effect of the use of particular textbooks. Finally, PISA and TIMSS provide the international perspective on achievement data for both primary and secondary school students.

Background information about the role of empirical research in mathematics education in the Netherlands can, for example, be found in Freudenthal (1977), Goffree (2002), Van den Brink and Streefland (1979), and Van der Velden (2000).

3.2   Students' Own Productions and Own Constructions—Adri Treffers' Contributions to Realistic Mathematics Education

Marc van Zanten[2]

3.2.1   Introduction

The development of Realistic Mathematics Education (RME) started with the setup of the Wiskobas project in 1968. Wiskobas is an acronym for 'Wiskunde op de basisschool', meaning mathematics in primary school. Treffers was one of the leading persons within Wiskobas from the beginning onwards. He can be considered as one of the founding fathers of RME. In an interview held on the occasion of the ICME13 conference, Treffers (Fig. 3.1) stated that, in his view, the active input of students in the teaching and learning process is the basis for good mathematics education. This was one of the issues that came to the fore in the interview, together with other ideas Treffers published on over the years, which are mentioned also in the text. In the interview, Treffers highlighted the issues that are most important to him, and looked back on his earliest sources of inspiration for his work on mathematics education. He explained how certain people had had a major influence on his vision of mathematics education. Interestingly, they turned out to be very special people not belonging to the

[2]Utrecht University & Netherlands Institute for Curriculum Development, the Netherlands, m.a.vanzanten@uu.nl

Fig. 3.1 Adri Treffers

community of mathematics education: he learned a lot about mathematics education from his five year older sister, his brother with whom he shared a bedroom, the father of his friend Beppie, and from Mr. Zwart, a very good teacher he had.

3.2.2   Treffers' Theoretical Framework for Realistic Mathematics Education

Treffers described in detail the principles of RME in his seminal publication *Three dimensions. A model of goal and theory description in mathematics instruction—The Wiskobas project* (1978, 1987a). The framework for an instruction theory for RME, formulated in the 1987 version, was built on the work of Wiskobas, established in collaboration with Freudenthal (1968, 1973), and his ideas about mathematics as a human activity, avoiding mathematics education as transmitting ready-made mathematics to students, and instead stimulating the process of mathematisation. This framework consisted of five instruction principles, derived from the didactical characteristics of Wiskobas. Over the decades these principles have been reformulated and further developed, both by Treffers himself and by other didacticians, and are still seen as the core teaching principles of RME, as described in the Sect. 3.1 of this chapter (see also Van den Heuvel-Panhuizen & Drijvers, 2014).

3.2.3   Students' Own Productions

According to Treffers (1987a, p. 249), teachers can help students to find their way to higher levels of understanding, but this "trip should be made by the pupil on his

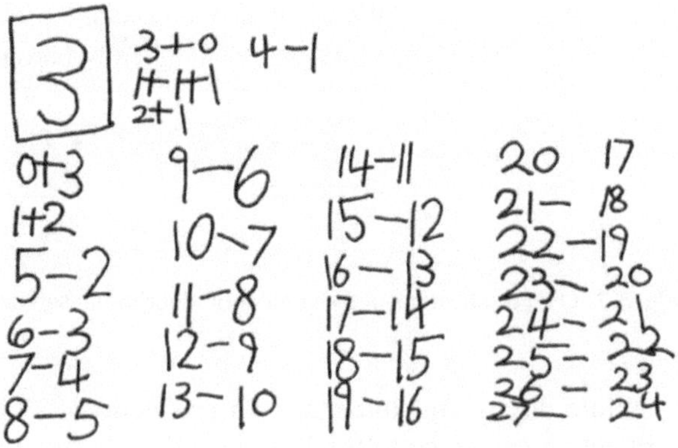

Fig. 3.2 An own production by a first grader (Grossman, 1975; see Treffers & De Moor, 1990, p. 163)

own two legs". Therefore, a decisive influence in the learning process comes from the students themselves in the form of their own productions and their own constructions. Referring to this contribution of the students to the learning process, Treffers (1987a, p. 250) speaks of an "essential factor", which explains his preference for students' own productions. In the interview Treffers revealed who formed the basis for this insight.

Treffers: My first source of inspiration is the most important one. That is my five year older sister. You could say that she invented a new didactic principle. Playing school in the attic, she let me and my friend Beppie make up our own mathematical problems. Nowadays we would call them students' own productions. An example of a completely free production is: "Produce problems that have 7 as the answer." To be honest, I don't think that we did it like that back then, but now we do.

Students producing problems themselves is one of the ways in which they can actively contribute to their own learning process. This can take place from the first grade on, as shown in Fig. 3.2. Here the assignment was to make up problems that should have 3 as the answer. These students' own productions can serve as 'productive practice', which can be done alongside to regular practice. One benefit of productive practice is that it engages many students. Another is that students are not limited to a certain range of numbers. As a result, students can actually surprise their teachers with their productions. For example, a student in the first grade who was asked to make up problems with the answer 5 came up with the problems '100–95', '2000–1995' and '10,000–9995', which were problems far beyond the number range the student had been taught at that moment. What happens when students make these productions is that they make use of the structure of the number system, which is a form of mathematisation.

Students' own productions evoke reflection, which stimulates the learning process. In particular, asking students to produce simple, moderate and complex problems can cause students to reflect on their learning path. Figure 3.3 shows an example of how students' own productions can be elicited in order to make them aware of what they find easy and difficult in algorithmic subtraction.

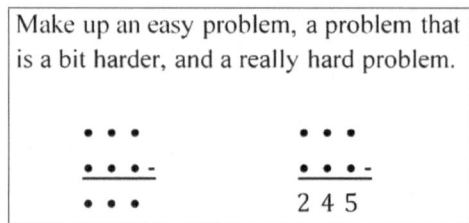

Fig. 3.3 Own productions in the context of algorithmic subtraction (Treffers, 2017, p. 85, p. 87)

While discussing students' own productions, all the difficulties of algorithmic subtraction can come to the fore, including borrowing once, borrowing more than once and borrowing from zero, Treffers points out. In line with this, Treffers comes up with another problem.

Treffers: Cover up some digits of a subtraction. Can another student reconstruct the original subtraction? How many digits can be covered up at max?

These questions may lead to more productive practice, but they also connect students' own productions to problem solving. The latter is significant because of the importance of challenging students, which came up later in the interview.

3.2.4 Students' Own Constructions

As mentioned previously, Treffers does not see the learning of mathematics as a process of absorbing ready-made knowledge. Instead, he considers the understanding of mathematics as a process that is constructed by students themselves.

Treffers: My second source of inspiration is my brother, with whom I shared a bedroom. He set me sometimes, teasingly, a few problems, like "you can't do those yet." But that taught me how to move along the imagined number line and flexible arithmetic, for example that you can calculate the multiplication tables in a smart way.

However, Treffers' view that students construct their own knowledge does not mean that mathematics education should rely on students' self-reliant discovery. Instead, instruction should make use of students' own contributions and should help them through 'guided reinvention', as Freudenthal called it. Students' own informal solution methods function as a starting point for such guided instruction. Figure 3.4 shows, as an example, informal constructions for 8 × 23 by students beginning to learn multi-digit multiplication.

In the various additive and multiplicative methods used by the students, several steps of the upcoming learning path are already recognizable. The teacher makes use of this in the interactive discussion of the students' solutions. He points out handy ways, such as in this case the use of products of the multiplication tables. In general, students are encouraged to think critically about their own solution methods

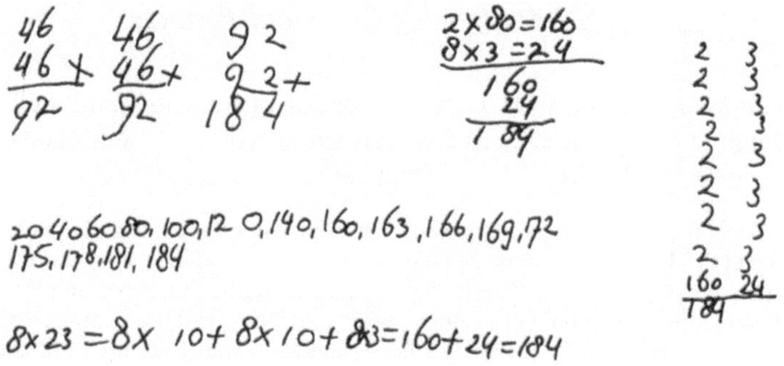

Fig. 3.4 Beginning third graders' constructions of solutions for 8 × 23 (Treffers, 1987b, p. 128)

Hans made a trip of 75 kilometer.
After 48 kilometer he took a brief rest.
1. How many kilometer must he do after the rest?
2. How do you think he travelled, by car, by bike, on foot, ...?
3. How long do you think the whole trip of 75 kilometer took him?

Fig. 3.5 The problem of Hans (Treffers, 1991, p. 338)

and compare them with their classmates' solutions and the solutions the teacher emphasizes. These latter are purposely chosen to guide the students to a gradual process of schematising, shortening and generalising.

The problem in Fig. 3.5 shows another type of problem that requires students' own constructions. Treffers (1991) is a strong proponent of these 'daily life' problems in which students have to make use of all kinds of measurement knowledge and have to figure out this knowledge based on their experiences with the situation involved. Moreover, this type of problems elicits reasoning and further questioning: How far is 75 km? Roughly how many km does a car cover in one hour? How fast does a bike go? What is the speed of a pedestrian? How long would a cyclist take to do 48 km? Taking a rest after half an hour, after two hours—what is sensible? What does a 'brief' rest mean? A quarter of an hour, half an hour, or a few hours? This kind of reasoning, involving arguing, proposing solutions and calculations in which knowledge of number and numerical data is both used and increased, is important for the development of students' numeracy. Treffers introduced this term and its importance in Dutch primary school mathematics education, leading to its inclusion in the officially established objectives for primary school.

3.2.5 Challenging Students with Classical Puzzles

Another feature of Treffers' work is the stimulation of students' thinking by offering them challenging problems, such as the subtraction problems mentioned earlier, but also more complicated, classical mathematical problems.

> Treffers: The father of my friend Beppie, a cobbler, gave us classical puzzles, including the 'Achilles and the Tortoise' paradox, and the famous 'Wheat and the Chessboard' problem. Later, a very good teacher, Mr. Zwart, elaborated these further, touching on problems we were still struggling with a bit. For example, whether 0,999… with the decimal 9 repeating infinitely is or is not equal to 1. Beppie's father and Mr. Zwart were my third and fourth sources of inspiration.

In his work, Treffers elaborated on the ways in which classical mathematical puzzles like the 'Wheat and the Chessboard' can be set in primary school so that there is, again, room for students' own constructions. Starting with students' informal approaches, discussing these in an interactive setting, leading the students to a shortened and structured procedure, all these features mentioned earlier are present in his descriptions. Due to the influential work of Treffers and his colleagues, these ideas made it into Dutch textbooks.

3.2.6 Students' Input Is the Basis of Everything

To conclude, Treffers comes with advice for mathematics education today.

> Treffers: I feel that of the things we spoke about, the issue of students' own productions is the most important one. Take for example, magic squares (Fig. 3.6). Students need to think about how to solve them, but they can also produce them for themselves. For the teachers, it means that they can see what students find easy or hard and how they can sometimes fly off far above the familiar range of number and the difficulty level of the problem they could have thought of themselves. Here you are talking about students who have an input in the teaching and learning process and that is the basis of everything.

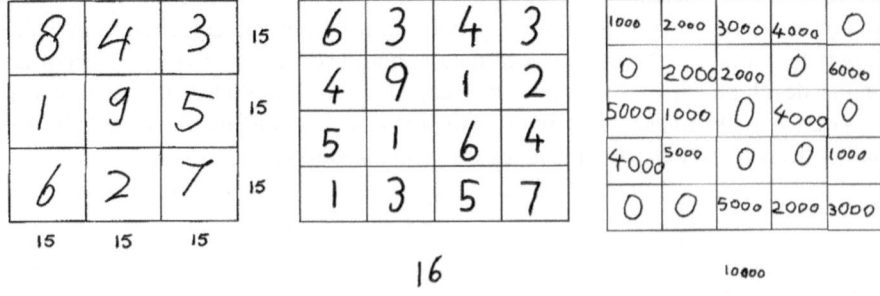

Fig. 3.6 Second-grade students' own productions of magic squares (De Goeij & Treffers, 2004)

Treffers sees students' own productions and constructions not only as a didactical tool, but also as a goal of mathematics education. Naturally this is true, since producing and constructing mathematics yourself is in essence mathematising. This ability is even more important in modern society than it already was in the early days of Wiskobas.

> Treffers: My general recommendation for the future of mathematics education is: enlarge the role of students' own productions and own constructions, in practice, in problem solving, and in the combination of the two.

3.3 Contexts to Make Mathematics Accessible and Relevant for Students—Jan de Lange's Contributions to Realistic Mathematics Education

Michiel Doorman[3]

3.3.1 Introduction

De Lange has been working at Utrecht University for 40 years. He led the Freudenthal Institute from 1981 and as Professor/Director from 1989 until 2005. He started as a mathematician and initially was more interested in upper secondary education. His most recent interests lie in the study of talents and competencies such as the scientific reasoning of very young children. De Lange worked on the theoretical basis of assessment design, carrying it through to practical impact in the Netherlands and internationally as chair of the PISA Mathematics Expert Group (Fig. 3.7).

De Lange's contributions to the ideas underpinning Realistic Mathematics Education (RME) are strongly connected to the role of contexts in mathematical problems. In traditional mathematics education, contexts are included in textbooks as word problems or as applications at the end of a chapter. These contexts play hardly any role in students' learning processes. Word problems are mostly short storylines presenting a mathematical problem that has a straightforward solution. Applications at the end of a chapter help students to experience how the acquired mathematical procedures can be applied in a context outside mathematics.

In RME, context problems have a more central role in students' learning process from the very start onwards. These problems are presented in a situation that can be experienced as realistic by the students and do not have a straightforward solution procedure. On the contrary, students are invited to mathematise the situation and to invent and create a solution. Ideally, students' intuitive and informal solutions anticipate the topics of the chapter and provide opportunities for the teacher to connect these topics to the students' current reasoning. In RME, such context problems are

---

[3] Utrecht University, the Netherlands, m.doorman@uu.nl

Fig. 3.7 Jan de Lange during the interview

expected to have a central role in the guided reinvention of mathematics by the students themselves.

This understanding of the potential of context problems was developed during the 1970s and 1980s and was largely inspired by the work of De Lange. His contributions to the Dutch didactic tradition consisted of developing a large collection of teaching units used mainly in innovation-oriented curriculum projects in the Netherlands and in the USA. In preparation for the Dutch contribution to ICME13 Thematic Afternoon session on European Didactic Traditions, De Lange was interviewed to reflect on his work and specifically on the importance of contexts in mathematics education.

3.3.2   Using a Central Context for Designing Education

After De Lange graduated in the 1970s he started his career as a mathematics teacher. Soon he discovered that students reacted quite differently to the topics that he tried to address in his lessons. Most surprising for him was that they did not recognize mathematics in the world around them. After De Lange moved to the Freudenthal Institute, one of his ambitions was to find contexts that could be used to make mathematics accessible. A teacher in lower secondary school asked him if he could do something for her students who had problems with trigonometric ratios. De Lange designed a unit intended for a couple of weeks of teaching.

> De Lange: I started to work on one of my hobbies, planes, and I wrote a little booklet. It is called *Flying Through Maths*. It is about all kinds of different mathematics, all in one context. It is about glide ratios, vectors and sine and cosine.

All the mathematics in this teaching unit is presented in the context of planes and flying, which approach is later referred to by De Lange as an example of 'central context design' (De Lange, 2015). This means that the same context is used to

---

GLIDE RATIO

The glide ratio is used to compare the performances of gliders or hang gliders. The first hang glider had a glide ratio of 1:4 (one to four). The improved version had a glide ratio of 1:7.

    4. Define what is meant by a glide ratio.

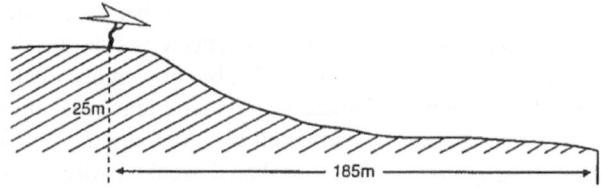

Otto Lilienthal made about 2500 flights. On one of his flights from the Rhinower Mountains near Berlin, he started from a height of 25 meters and flew a distance of 185 meters. On his next flight he changed his glider a bit, and then started at a height of 20 meters to reach 155 meters.

    5. What were the glide ratios of Otto Lilienthal's two gliders? Was the second glider better than the first one?

Fig. 3.8   *Glider* problem from the booklet *Flying Through Maths* (De Lange, 1991, p. 7)

introduce students into various mathematical concepts. One of the concepts that was presented in this flying context was the glide ratio (Fig. 3.8).

The context of flying is used to encourage students to reason about covering distances when gliding from a certain height. By comparing different flights, students are expected to come up with some thinking about the glide ratio, i.e., the ratio between the distance covered and the starting height. This glide ratio plays a role in the context problems in the beginning of a chapter on slopes. In this way, the glide ratio is meaningful for students. They can use it for solving problems that they can experience as real problems. Later in the chapter this glide ratio is generalized to triangles and as a measure that can be used to calculate or compare slopes.

3.3.3   Contexts for Introducing and Developing Concepts

RME brought about a new perspective on the use of contexts in mathematics education. Contexts are not only considered as an area for application learned mathematics, but also have an important role in the introduction and development of mathematical concepts.

> De Lange: Applications is one thing. In the traditional textbooks, it was the end of the book. You started first with learning mathematics, and then you got the applications of mathematics. Through our theory developed at the Freudenthal Institute in the 70s, we changed that to developing concepts through context. So, you had to be very careful, because if the context is not very suited for the concept development, you are riding the wrong train on the wrong

track. But I think in general we can say for a lot of concepts we found very nice contexts to start with.

An example is the context of exponential growth that can support students in developing the logarithm-concept (De Lange, 1987). In this context, the growth of water plants, students first calculate the exponential increase of the area covered by these water plants with the growth factor and the number of growing weeks. At a certain moment, the question in the context is reflected. The question is no longer what the area is after a certain number of weeks, but how many weeks are needed to get an area that is 10 times as much? Students will experience that this is independent of the starting situation and can be estimated by repeating the growth factor (e.g., with a growth factor of 2 this is a bit more than 3 weeks). After these introductory tasks, $^2\log 10$ is defined as the time needed to get 10 times the area of water plants when the growth factor per week is 2. With this context and the concrete contextual language in mind, students can develop basic characteristics of logarithmic relations such as $^2\log 3 + 1 = {}^2\log 6$ as follows: with this 1 extra week, you get 2 times more than 3, which equals 6. Similarly, $^2\log 6 + {}^2\log 2$ has to equal $^2\log 12$, as $^2\log 6$ is the number of weeks to get 6 times as much, and $^2\log 2$ is the number of weeks to get 2 times as much. The time needed to first get 6 times as much followed by the time needed to get 2 times as much has to be equal to the time needed to get 12 times as much.

With such a context problem, a concept is not only explored, but also more or less formalized. Such a concrete foundation is important because it offers opportunities for students in the future to reconstruct the procedure and meaning of the abstract calculation procedures by themselves.

3.3.4   Relevant Mathematics Education

The aforementioned examples show the potential of contexts for learning mathematics and for making that learning process meaningful and relevant for students. This approach to mathematics education connects to the RME instructional theory in which the learning of mathematics is interpreted as extending your common sense reasoning about the world around you. Hence, De Lange emphasises in his reflection on educational design that designers need to find contexts by meeting the real world outside mathematics and experience the potential of contexts by going to real classrooms (De Lange, 2015). Observing authentic classroom activities is crucial. He stresses the importance of direct observations without using video in order to observe much more. In such a direct observation, one is able to look at the students' notes, one has the possibility to participate with their work, and one can ask questions in order to understand why students do what they do.

Exploiting the real world guided De lange towards a wide variety of original and surprising contexts. One example is the art of ballooning. Flying a balloon depends completely on the strength and direction of the wind and the change of the wind

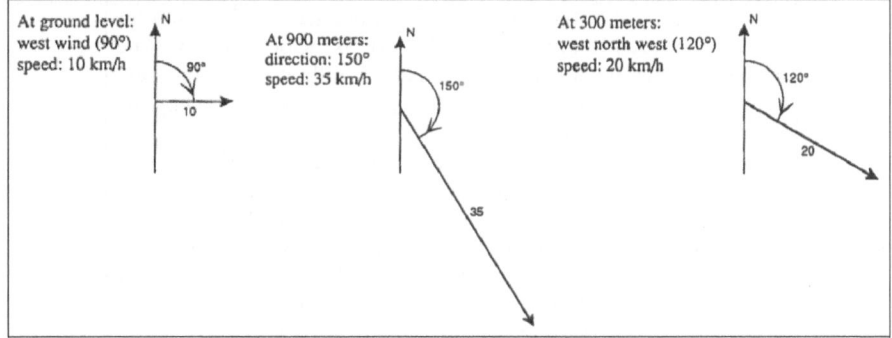

Fig. 3.9 Ballooning information from the booklet *Flying Through Maths* (De Lange, 1991, p. 34)

> In the Netherlands, one out of four high school students has used drugs (at least once). Design a study to find out how many students at your school have used drugs. Do you get a representative sample by using your school when you want to say something about the use of drugs at all schools in your city?

Fig. 3.10 *Drug use* problem (De Lange & Verhage, 1992, p. 12)

speed along different altitudes. The following example (Fig. 3.9) is also taken from the booklet *Flying Through Math* and is typical for a situation in which you are supposed to travel with a balloon from one spot to a target. The task for the students was to determine what happens when a balloon starts from Albuquerque and flies the first half hour at 300 m, then an hour at 900 m and finally, a half hour at 300 m.

In this context, not all information is available. Remaining questions are: How much time is needed to land? What happens when you go from one altitude to another and how much time does that take? The task becomes a real problem solving task for the students.

Through being in real classrooms De Lange could also look for contexts that trigger interest in students. In choosing these contexts, he was not afraid of using controversial situations. This can be recognized in a task for 15-year olds about drug use (see Fig. 3.10), for example.

In the interview, De Lange emphasised again that contexts serve many important roles in the teaching and learning of mathematics. They support conceptual development, can be motivating and raise interest, and also teach students how to apply mathematics.

> De Lange: We should be aware that contexts have to be mathematised. This means that we should be aware of what is the relevant mathematics in the contexts, which concepts plays an important role, and can the contexts serve as the starting point of modelling cycles. So, what you actually see is, that in the first phase of learning from context to concept, you use things, you do things, which are exactly the same as using the concept in a problem-solving activity.

In a certain country, the national defence budget is $30 million for 1980. The total budget for that year is $500 million. The following year the defence budget is $35 million, while the total budget is $605 million. Inflation during the period covered by the two budgets was 10 percent.

 a. You are invited to give a lecture for a pacifist society. You intend to explain that the defence budget decreased over this period. Explain how you could do this.

 b. You are invited to lecture to a military academy. You intend to explain that the defence budget increased over this period. Explain how you would do this.

Fig. 3.11 *Military-budget* problem (De Lange, 1987, p. 87)

The examples of problems discussed above all illustrate how contexts can be used for designing relevant and meaningful mathematics education. What can also be recognized is that in the 1980s designers like De Lange already anticipated what we now call 21st century skills. A nice example of this is the *Military-budget* problem (Fig. 3.11), which some thirty years ago was designed by De Lange to stimulate students to become mathematically creative and critical.

3.3.5 Conclusion

Creative designers like De Lange, people who are able to convince others of the limitations of many textbooks and who are able to translate general educational ideas into original and attractive resources for students, are of crucial importance for realising meaningful and relevant mathematics education. In 2011, he was awarded the *ISDDE Prize for Excellence in Design for Education*. Malcolm Swan wrote on behalf of the prize committee: "He has a flair for finding fresh, beautiful, original, contexts for students and shows humour in communicating them." Without De Lange's contributions many ideas in the Dutch didactic tradition would have been less well articulated, less well illustrated, and less influential in the world outside the Dutch context.

3.4 Travelling to Hamburg

Paul Drijvers[4]

3.4.1 Introduction

This section describes an example of a task that was designed and field-tested for the ICME13 conference. Its aim is to illustrate how the principles of Realistic Mathematics Education (RME) (see, e.g., Van den Heuvel-Panhuizen & Drijvers, 2014). can guide the design of a new task. Indeed, task design is a core element in setting up mathematics education according to an RME approach. This is one of the reasons why design-based research is an important research methodology in many studies on RME (Bakker & Van Eerde, 2015).

The task presented here involves setting up a graph that many students are not familiar with, because it displays one distance plotted against another. For several reasons, it makes sense to have students work on such less common graphs. First, graphs in mathematics education in almost all cases involve an independent variable, often called x, on the horizontal axis, and a dependent variable, for example y or $f(x)$, on the vertical axis. However, there are also other types of graphs than these common x-y graphs. In economics, the independent variable—not always x but also t for time—can also be plotted on the vertical axis rather than on the horizontal. In physics—think about phase diagrams—the independent variable may be a parameter that is not plotted on one of the axes. The latter case reflects the mathematical notion of parametric curve, in which the independent variable remains implicit. In short, students should be prepared for other types of graphs as well.

A second, more general reason to address non-typical types of graphs is the world-wide call for mathematical thinking and problem solving as overarching goals in mathematics education (Devlin, 2012; Doorman et al., 2007; Schoenfeld, 1992). If students are to be educated to become literate citizens and versatile professionals, they should be trained to deal with uncommon problem situations that invite flexibility. As such, mathematical thinking has become a core aim in the recent curriculum reform in the Netherlands (Drijvers, 2015; Drijvers, De Haan, & Doorman, submitted).

In this section, first the task will be presented together with some design considerations that led to its present form. Next, a brief sketch is provided of the results of the field test in school and of what the task brought to the fore at the ICME13 conference. As a task may need adaptation to the specific context in which it is used, its elaboration is described next for the purpose of in-service teacher training, including a three-dimensional perspective. Finally, the main points will be revisited in the conclusion section.

---

[4]Utrecht University, the Netherlands, p.drijvers@uu.nl

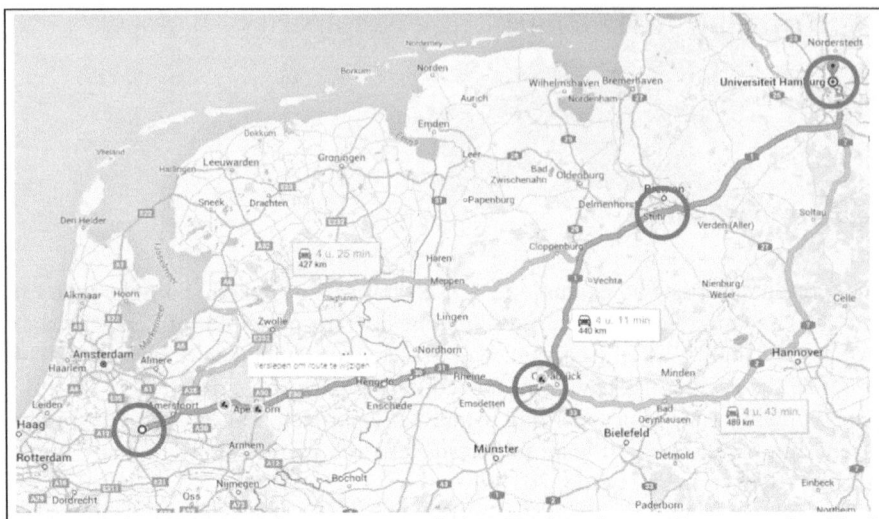

To attend the ICME13 conference, we travelled to Hamburg by car. To prepare for this, I consulted Google Maps. Here you see the result: we leave Utrecht, in the Netherlands. The distance to Hamburg is 440 km. After 215 km, we pass by Osnabrück. Another 115 and we pass by Bremen. Next, we need to drive the final 110 km to reach Hamburg.

Fig. 3.12 Setting the scene for the task

3.4.2   Task Design

Figure 3.12 shows the presentation of the task in the form of a picture, displayed through a data projector, and a suggested text that might be spoken by the teacher. As the problem situation is a somewhat personal story from 'real life', it is preferable to deliver the text orally rather than in written form. It is expected that the task becomes 'experientially real' through this form of presentation. In the task, the perspective taken is that of a participant in the ICME13 conference in Hamburg, Germany. Of course, this perspective could easily be adapted to other situations that are more relevant to the audience.

Figure 3.13 shows how the task presentation might continue. It shows a schematisation of the problem situation, in which the 'noise' of the real map has been removed. In the text that might be spoken, this schematisation and the underlying mental step of representing the highway as a line segment are explicitly addressed. Depending on the audience and the intended goal of the task, of course, one might consider leaving this step up to the student and to reduce guidance at the benefit of opportunities for guided reinvention. For the field tests addressed in the next section, it was decided not to do so, due to the expected level of the students and the time constraints. The text in Fig. 3.13 ends with the problem statement. Students are invited to use their worksheet, which contained two coordinate systems like the one shown in Fig. 3.14.

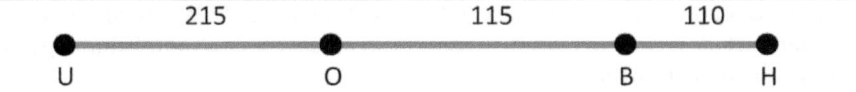

215	115	110

U   O   B   H

If I depict the highways as straight-line segments, I can globally represent the situation in this way. U stands for Utrecht, 215 to O (Osnabrück), 115 to B (Bremen), and 110 to H (Hamburg). Now I am wondering the following. During the trip, while driving, my distances to both Osnabrück and Bremen constantly change. What would a graph look like, if I put the distance to Bremen vertically, and my distance to Osnabrück horizontally? Could you please take a few minutes to sketch the graph in which the distance to B during the trip is vertically plotted against the distance to O on the horizontal axis? Please use your worksheet!

Fig. 3.13 Schematising the situation and posing the problem

Fig. 3.14 Coordinate system that is shown on the worksheet

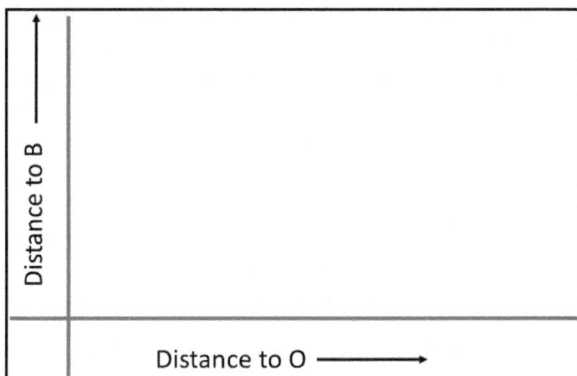

One might wonder if this is a realistic task. Who would be so silly as to raise this question? Well, we were. We were three mathematicians who were bored during the long trip to Hamburg. How to "sell" this to students? A possible approach could be to "play the card of the strange mathematician", but this should not be exaggerated. Our experience is that students may be intrigued by such problem situations, even if the question itself is not solving a 'real' problem. Also, even if one might ask "Why do we need to know this?", the problem situation is sound in the mathematical sense, and has possibilities for applications in mathematics and science, as explained in Sect. 3.4.1.

After the task is presented, students can start to work in pairs, in small groups, or individually. As it is expected that there will be quite a bit of differentiation in the class—some students might solve this task in a minute, whereas others may not have a clue where to start—the students who finish quickly can be provided orally with an additional task:

If you feel you are doing well, please think of a question that you might use to help a peer who doesn't know how to start, a question that might serve as a scaffold.

Also, some scaffolding questions are prepared that may serve as a hint to react to students who have difficulties with the task and raise their hands for help. For example: How can you make a start? Do you know a similar but easier problem?

Does this resemble a problem that you have seen in the past? Where are you in the *O-B* plane when you are leaving Utrecht? And when you arrive in Hamburg? And when you pass by Osnabrück?

To be effective, such a class activity needs a whole-class wrap-up. It might start with the question of how to help peers to make a start, or how you started yourself. For example, one can consider finding the position in the plane for the special moments of leaving Utrecht and arriving in Hamburg. This leads to the points $(O, B) = (215, 330)$ and $(O, B) = (225, 110)$. How about, when passing Osnabrück and Bremen? Another option is to imagine what happens in between O and B: the distance to O increases as much as the distance to B decreases. How does this affect the graph? A natural question that emerges, is whether the driving speed should be constant, and if it matters at all. Would the graph look different if you walk from Utrecht to Hamburg rather than driving (not recommended, of course)? In an advanced class, with many students coming up with a sensible graph, it might be interesting to show an animation in a dynamic geometry environment, which in its turn may invite setting up parametric equations. Of course, how far one can go in such a wrap-up largely depends on the students' progress. If needed, postponing the presentation of the results to the next lesson may be an appropriate 'cliff-hanger teaching strategy'. The expertise and the experience of the teacher in leading the whole-class wrap-up are decisive in making the task work in class.

In retrospective, the following considerations guided the design of this task:

- To make the problem situation come alive for the audience at the ICME conference, the trip to Hamburg was chosen as a point of departure. The 'experientially real' criterion was decisive.
- In the beginning, there was some hesitation on whether to travel by train or by car. The advantage of the train would have been that, contrary to cars driving on highways, trains do pass through the city centres. However, it was estimated that the car version would be more recognizable to the audience, particularly in combination with the Google Maps image and driving directions. As an aside, the designers of the tasks did not travel by car to Hamburg themselves; the story is based on another car trip. The point in designing this task is not the truth of the story behind it, but its experiential reality and mathematical soundness.
- How openly to phrase the problem? When designing the problem different versions came up, with different levels of support. Indeed, the version we had in mind might be quite a surprising challenge to students, but it was expected that through the scaffolding hints mentioned above, it would be possible to have the students start.
- How to present the problem? It was decided to present the task orally to the class as a whole, supported by slides displayed through a data projector. The idea here was that this would enhance the personal character of the problem situation. Also, such an oral whole-class introduction is expected to provide a collaborative setting, while working on a shared problem. Finally, an oral presentation can be a welcome change after many textbook-driven activities.

- It was decided to provide the students with the crucial linear representation (Fig. 3.13) and, in this way, give away the first schematisation step. Other choices can make perfect sense here. All depends on the level of the students, their preliminary knowledge, the time available, and the learning goals.
- To deal with student differences in this task, a second layer was built in, namely, that of thinking of hints for peers. In this way, students who finished the task quickly were invited to put themselves in the place of their slower peers, and, as a consequence, reflect on the thinking process needed to solve the task.

3.4.3 Field Tests

To prepare the activity for the ICME13 conference, the task was field-tested in a bilingual class in a rural school in the Netherlands. The students, 13- to 14-year olds, took part in the pre-university stream within secondary education. The pilot took one 50-minute lesson. After the oral introduction, students went to work. The question needed to be repeated once or twice. Also, we had a short whole-class discussion after the first tentative graphs, and invited the students to sketch a second one afterwards.

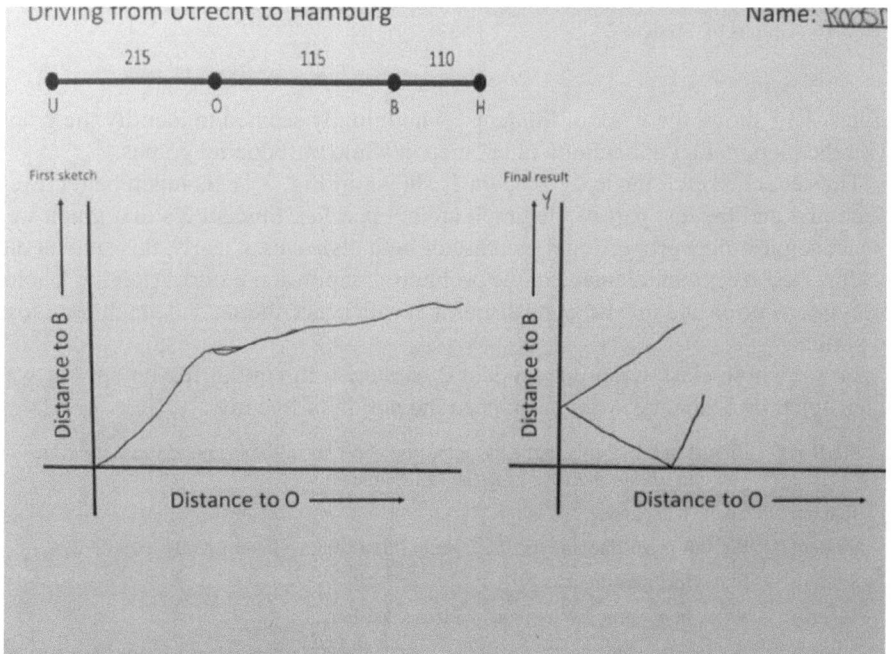

Fig. 3.15 Graphs by Student 1

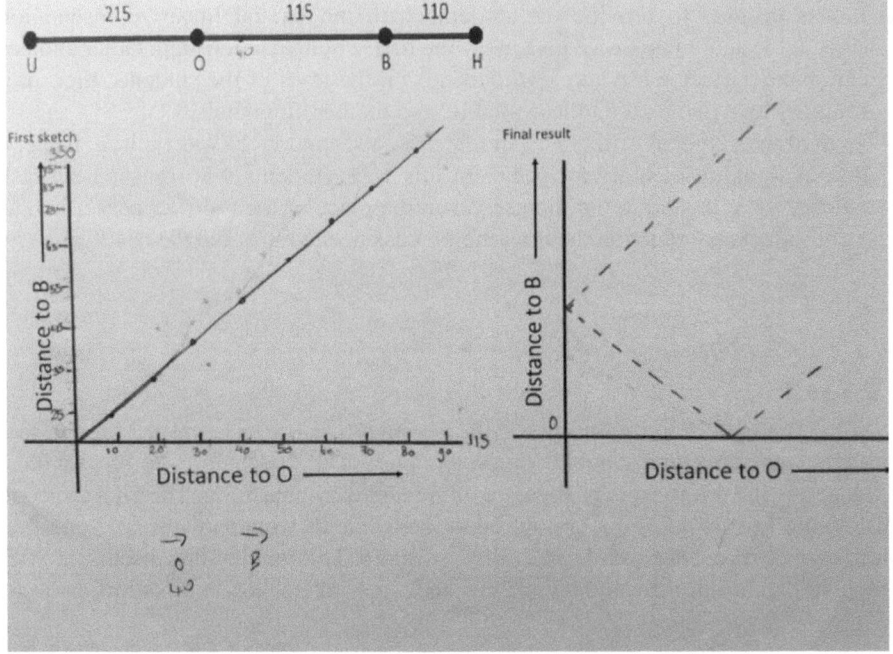

Fig. 3.16 Graphs by Student 2

Figure 3.15 shows the work of Student 1 who initially seemed to identify the graph with the map, not an uncommon phenomenon while introducing graphs.

The second sketch made by Student 2, shown in Fig. 3.16, is much better, even if the first and the last part of the graph are not parallel. Student 2's first graph was linear, suggesting a proportional increase of both distances. Clearly, this student did not have a correct mental image of the problem situation at the start. After the whole-class interruption and maybe some discussion with peers, the second graph was close to perfect.

In the whole-class wrap-up, Student 2 explained his initial reasoning, but was interrupted by Student 3, who introduced the notion of linearity.

Student 2: First, I had like this, but I thought, you can't be in the origin at the same time, you can't be in *B* and in *O* at the same time ….

Teacher: Yeah, you cannot.

Student 2: So, I thought like, maybe they, yeah, I don't know, I can't really explain it.

Student 3: It's a linear formula.

Teacher: Wow, how come? Why is it…. Please, explain.

Student 3: Well, ehm, since there isn't, eh yes, since the amount added always is the same, the first step, it's a linear formula.

This short one-lesson intervention confirmed the initial expectations, that the problem situation was rich and could give rise to interesting discussions.

During the ICME13 conference in Hamburg, the task was piloted again in a similar way with an audience of about 250 attendants. Of course, individual help was hard to deliver in this large-scale setting. Still, in comparison to the field test in class, similar patterns could be observed. Also, the need for level differentiation was bigger than in the secondary class, due to the heterogeneity of the ICME audience. It was surprising how mathematics teachers, researchers and educators have their schemes for graphing, and can get quite confused once these schemes are challenged by new situations.

3.4.4 Possible Task Extensions

As already mentioned, guided reinvention, meaning and experiential reality are subtle matters. To be able to deal with this subtlety appropriately in a setting with students of different levels, a good task should provide teachers with opportunities to simplify the task, to provide variations, and to deepen and extend the task. A straightforward way to simplify the task is to leave out one of the two cities between Utrecht and Hamburg, or even to leave out both and ask for the graph of the distance to Hamburg against the distance to Utrecht. These might be appropriate first steps towards solving the original problem. As variations, one may consider similar situations, such as the already mentioned trip by train, or a bike ride from home to school.

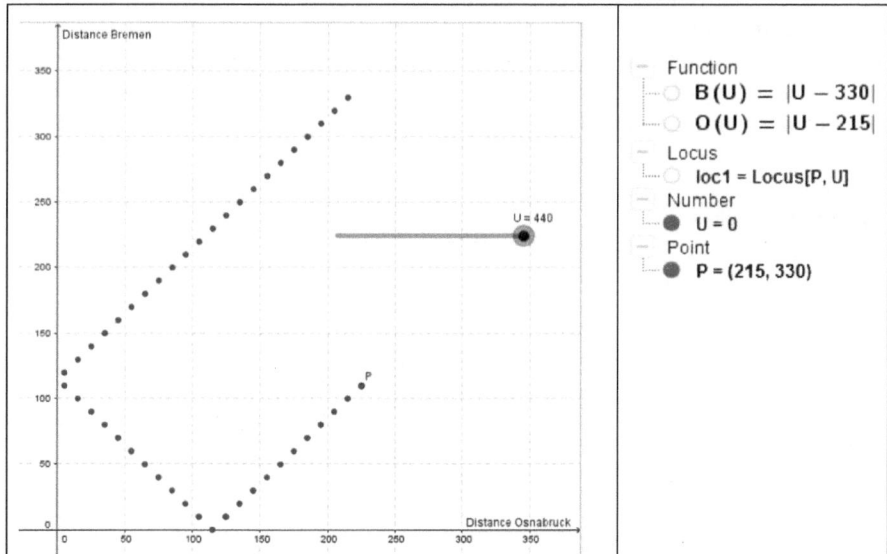

Fig. 3.17 Animation in Geogebra (left) and the underlying function definitions (right)

As students may come up with different graphs and will make all kinds of gestures while explaining their reasoning, it might be convenient to show the graph through an animation in a dynamic geometry system. The left screen of Fig. 3.17 shows such an example in Geogebra, using a slider bar to move the point. This may help to illustrate the resulting graph. In the meantime, however, this raises a deeper question: How can you make this animation, which equations and definitions are needed? The right screen in Fig. 3.17 provides the answer. The following definitions were used:

• Distance to Utrecht: U	(Independent variable)
• Distance to Osnabrück: $O(U) = |U - 215|$	(Dependent variable)
• Distance to Bremen $B(U) = |U - 330|$	(Dependent variable)
• Point in the plane: $P(U) = (O(U), B(U))$	

In this way, we take a mathematical perspective and the problem forms a gateway to the fascinating world of parametric curves.

As a final extension, also a third city between Utrecht and Hamburg can be considered. For example, Cloppenburg is about in the middle of Osnabrück and Bremen: Osnabrück–Cloppenburg is 60 km, and Cloppenburg–Bremen is 55 km. Can you plot a graph, indicating how the distances to Osnabruck, Cloppenburg and Bremen co-vary during the trip? Note that this task, in line with its higher level, is phrased in a somewhat more abstract way. Of course, the graph in this case will be in three dimensions rather than in two. Again, an animation can be built in Geogebra (Fig. 3.18). Rotating the graph shows a familiar form (Fig. 3.19) and in a natural way raises new, interesting questions, such as on the angle between the trajectory and the planes. This latter extension to the third dimension was used in a teacher professional development course, in which the participants found the two-dimensional case relatively easy, but were intrigued by the problem situation.

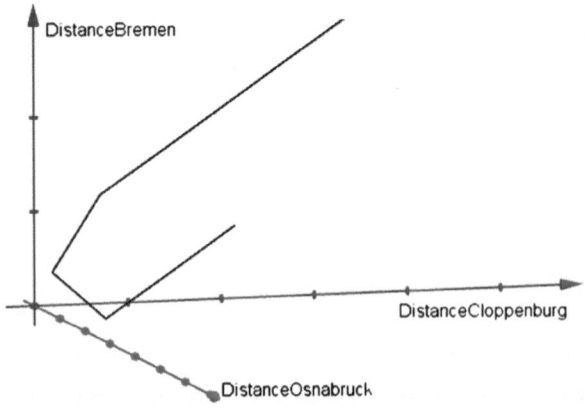

Fig. 3.18 3D graph in Geogebra

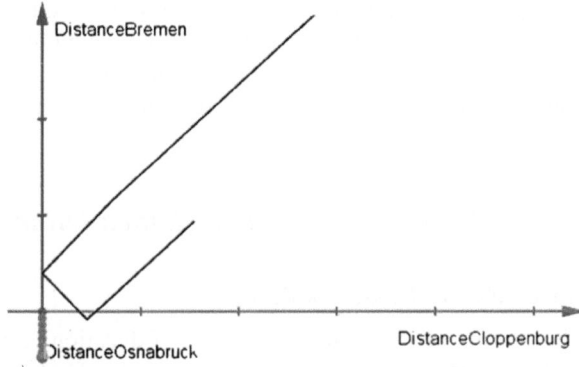

Fig. 3.19 3D graph in Geogebra seen from one of the axes

3.4.5   *Conclusion*

This example on task design according to RME principles revealed that for both students in Dutch secondary school and for participants of the ICME13 conference, it was hard to have the flexibility to refrain from the conventional time-distance graph paradigm and to open the horizon towards distance-distance graphs. This type of mathematical flexibility, needed in this unconventional and non-routine task, is core in problem solving, and at the heart of what RME sees as an essential value in mathematics education. The point of departure is 'realistic' in the sense that both target groups could imagine the situation and seemed to perceive it as realistic. What makes the task suitable from an RME perspective is that it can be used in different variations, appropriate for different levels of students and for different mathematical learning goals. Also, there are different, more and less mathematical, approaches and solution strategies, as well as follow-up questions. Finally, the somewhat surprising character of the task, may lead to the kind of lively and mathematically interesting interactions among students and between students and their teacher that are so important in the co-construction of mathematical meaning.

These task characteristics are central in RME and reflect the approach to mathematics education in the Dutch didactic tradition. To design tasks that elicit genuine mathematical activity in students is a challenge, not only in the Netherlands but in the mathematics education community world-wide!

3.5   Voices from Abroad

The chapter concludes with five sections which give a flavour of the international life of RME. From the beginning of the development of RME, mathematics educators all over the world were interested in it. This led to cooperation with a large number of countries where RME ideas and materials were tried out, discussed and adapted.

The countries that are represented here in these short notes about their experiences with RME all had a prominent place in a certain phase of the development of RME, and cover all corners of the globe, including the United States, Indonesia, England and the Cayman Islands, South Africa, and Belgium.

3.5.1   Realistic Mathematics Education in the United States

David Webb,[5] **Frederick Peck**[6]

The origins of Realistic Mathematics Education (RME) in the United States can be traced back to a proof-of-concept study (De Lange, Burrill, Romberg, & van Reeuwijk, 1993) at a high school in Milwaukee organized by Romberg (University of Wisconsin) and De Lange (Freudenthal Institute). The success of this pilot study illustrated how RME design principles could be applied in U.S. classrooms. More recently, RME continues to be articulated largely through professional development opportunities offered at innovation centres. As we trace the spread and scale of RME in the United States, the instantiation of RME is best characterised as a teacher-centred approach that involves principled reconsideration of how students learn mathematics. Reconsideration of beliefs and conceptions is often motivated when teachers re-experience mathematics through the lens of progressive formalisation and related didactic approaches (Webb, Van der Kooij, & Geist, 2011). Teachers' participation in the interpretation and application of RME in U.S. classrooms has led to systemic innovation that has been sustained, inspired and supported by professional development and curricula, and by fellow teachers who provide their colleagues with a proof-of-concept in their local context.

3.5.2   Two Decades of Realistic Mathematics Education in Indonesia

Zulkardi,[7] **Ratu Ilma Indra Putri,**[8] **Aryadi Wijaya**[9]

In Indonesia, some two decades ago, the process of adapting Realistic Mathematics Education (RME) began (Sembiring, Hadi, & Dolk, 2008). The Indonesian approach is called 'Pendidikan Matematika Realistik Indonesia' (PMRI). This process began in 1994 when Sembiring from the Institut Teknologi Bandung met De Lange, the director of the Freudenthal Institute of Utrecht University, who was presenting a keynote

---

[5]University of Colorado Boulder, United States, dcwebb@colorado.edu

[6]University of Montana, United States, frederick.peck@mso.umt.edu

[7]Sriwijaya University, Indonesia, zulkardi@yahoo.com

[8]Sriwijaya University, Indonesia, ratu.ilma@yahoo.com

[9]Yogyakarta State University, Indonesia, a.wijaya@uny.ac.id

at the ICMI conference in Shanghai. The next step was the decision of the Indonesian government to send six doctoral candidates to the Netherlands to learn about RME. Afterwards the development and implementation of RME was continued through a Dutch-Indonesian project called 'Do-PMRI' (Dissemination of PMRI). Moreover, implementation strategies were carried out such as developing a master's program on RME, designing learning materials using RME theory and the development of a national contest of mathematics literacy using context-based mathematics tasks similar to those employed in the PISA test (Stacey et al., 2015). Recently, there were two new initiatives at Sriwijaya University in Palembang, namely the development of a Centre of Excellence of PMRI and the establishment of a doctoral programme on PMRI.

3.5.3 Implementing Realistic Mathematics Education in England and the Cayman Islands

Paul Dickinson,[10] **Frank Eade,**[11] **Steve Gough,**[12] **Sue Hough,**[13] **Yvette Solomon**[14]

Realistic Mathematics Education (RME) has been implemented in various projects over the past ten years in the secondary and post-16 sectors of the English education system. All of these projects can be characterised as needing to deal with clashing educational ideologies. In particular, pressure towards early formalisation and the heavy use of summative assessment has influenced how far it is possible to change teachers' practices and classroom cultures. Nevertheless, intervention studies based on the use of RME materials and RME-inspired pedagogic design showed that, in post-tests, students were willing to 'have a go' at problems, indicating confidence in their ability to make sense of a problem and to apply their mathematics in different contexts. Also, students were able to use a range of strategies to answer questions, including a use of models which reflected higher levels of sense-making in mathematics than before (Dickinson, Eade, Gough, & Hough, 2010). Even in post-16 national examination resit classes with students who had experienced long-term failure in mathematics, and where teaching is normally focused on examination training, small but significant achievement gains were found in number skills following a short intervention (Hough, Solomon, Dickinson, & Gough, 2017). While questionnaire data did not show any improvement in students' attitudes to mathematics, analysis of interview data suggested that this finding reflected the long-term legacy of their previous experience of learning mathematics by learning rules without meaning, but nevertheless, many students reported enjoying mathematics more in RME classes. In

[10]Manchester Metropolitan University, United Kingdom, p.dickinson@mmu.ac.uk

[11]Manchester Metropolitan University, United Kingdom, frankeade@outlook.com

[12]Manchester Metropolitan University, United Kingdom, s.j.gough@mmu.ac.uk

[13]Manchester Metropolitan University, United Kingdom, s.hough@mmu.ac.uk

[14]Manchester Metropolitan University, United Kingdom, y.solomon@mmu.ac.uk

the Cayman Islands, with an education system that is influenced by British tradition, but which is distant from many of its politically driven accountability pressures and measures, the RME approach with primary students who had poor number sense led to a substantial gain in achievement. Also in secondary school, students were very positive about RME materials and made improvements in achievement as well. Teachers agreed to continue using contexts, interactive approaches and models to support problem solving rather than focus on formal algorithms. So, despite the problems encountered in these projects, there are reasons to remain optimistic about the potential of an RME approach in the English system.

3.5.4   Reflections on Realistic Mathematics Education in South Africa

Cyril Julie,[15] Faaiz Gierdien[16]

The project Realistic Mathematics Education in South Africa (REMESA) was introduced in South Africa during a period when curriculum changes were introduced to fit the educational ideals of the 'new' South Africa. In the project, a team comprising staff from the Freudenthal Institute and the Mathematics Education sector of the University of the Western Cape develop several RME-based modules, which were implemented in classrooms. One of the modules was *Vision Geometry* (Lewis, 1994). It was deemed that this topic was a sound way to manifest the RME approach. Content of vision geometry such as lines of sight, angle of sight, perspective, was encapsulated in activities for students. Although the students found the activities enjoyable and not above their abilities, the teachers had concerns about the time needed for the activities, the curriculum coverage and the examinability of the module's content. This scepticism remained after the module was somewhat adapted. Another module that was developed was *Global Graphs* (Julie et al., 1999). This module was also adapted by the South African staff. For example, another introduction was chosen than in the original RME version, namely instead of having the students construct graphs, asking them to match graphs and situations. A difference with the other module was that it was developed with a larger group of practising teachers. Moreover, the teaching experiments occurred when there was a more stable, albeit contested, operative curriculum in the country. Overall, teachers expressed satisfaction about the usefulness of the module *Global Graphs*. In contrast with the module *Vision Geometry*, the module *Global Graphs* was more readily accepted, which also can be ascribed to the prominence of graphical representations in South African school mathematics curricula. The lesson learned from the REMESA project is that the proximity of innovative approaches to the operative curriculum plays an important role in teachers' adoption of resources for their practice. In addition, the REMESA

---

[15]University of the Western Cape, South Africa, cjulie@uwc.ac.za

[16]University of Stellenbosch, Stellenbosch, South Africa, faaiz@sun.ac.za

project has contributed positively to current research and development endeavours to address the issue of high-quality teaching of mathematics in secondary schools in low socio-economic environments in a region in South Africa.

3.5.5   Influences of Realistic Mathematics Education on Mathematics Education in Belgium

Dirk De Bock,[17] **Wim Van Dooren,**[18] **Lieven Verschaffel,**[19] **Johan Deprez**[20]**, Dirk Janssens**[21]

The second half of the last century was a turbulent time for mathematics education in Belgium. In the 1960s and 1970s, mathematics education—as in many other countries—was drastically changed by the New Math movement that broke through (Noël, 1993). This revolution first took place in secondary school and entered primary school a few years later. Then, for about twenty years, the official curricula in Belgium followed this New Math approach faithfully. When from the 1980s on, New Math was increasingly criticised in Flanders, it was opted for a reform along the lines of Realistic Mathematics Education (RME) (De Bock, Janssens, & Verschaffel, 2004). In the end, this led to a reformed Flemish primary school curriculum that indeed is strongly inspired by the Dutch RME model, but that certainly is not simply a copy of that model. Although, for example, more attention was paid to linking numbers to quantities and to solution methods based on heuristic strategies in addition to the standard computational algorithms, the Flemish curriculum maintained the valuable elements of the strong Belgian tradition in developing students' mental and written calculation skills. Also in secondary school the critique of New Math led to using elements of the Dutch RME model to enrich the Belgian mathematics curriculum. Among other things, this resulted in giving solid geometry a more prominent place, making geometry more connected to measurement, and having a less formal, intuitive-graphical way of introducing calculus. In addition to other content changes, this reform brought also a number of didactic innovations, such as the role given to modelling and applications and to authentic mathematical exploration, discovery and simulation.

---

[17]KU Leuven, Belgium, dirk.debock@kuleuven.be

[18]KU Leuven, Belgium, wim.vandooren@kuleuven.be

[19]KU Leuven, Belgium, lieven.verschaffel@kuleuven.be

[20]KU Leuven, Belgium, johan.deprez@kuleuven.be

[21]KU Leuven, Belgium, dirk.janssens1@kuleuven.be

References

Bakker, A. (2004). *Design research in statistics education: On symbolizing and computer tools.* Utrecht, The Netherlands: CD-Bèta Press/Freudenthal Institute, Utrecht University.

Bakker, A., & Van Eerde, D. (2015). An introduction to design-based research with an example from statistics education. In A. Bikner-Ahsbahs, C. Knipping, & N. Presmeg (Eds.), *Doing qualitative research: Methodology and methods in mathematics education* (pp. 429–466). Berlin: Springer.

De Bock, D., Janssens, D., & Verschaffel, L. (2004). Wiskundeonderwijs in Vlaanderen: Van modern naar realistisch? [Mathematics education in Flanders: From modern to realistic?]. In M. D'hoker & M. Depaepe (Eds.), *Op eigen vleugels. Liber Amicorum Prof. dr. An Hermans* (pp. 157–169). Antwerpen-Apeldoorn: Garant.

De Goeij, E., & Treffers, A. (2004). Tovervierkanten [Magic squares]. *Willem Bartjens, 23*(3), 28–32.

De Lange, J. (1987). *Mathematics, insight and meaning.* Utrecht, The Netherlands: OW & OC, Utrecht University.

De Lange, J. (1991). *Flying through maths.* Scotts Valley, CA: Wings for Learning.

De Lange, J. (2015). There is, probably, no need for this presentation. In A. Watson & M. Ohtani (Eds.), *Task design in mathematics education—ICMI Study 22* (pp. 287–308). Cham: Springer.

De Lange, J., & Verhage, H. (1992). *Data visualisation.* Scotts Valley, CA: Wings for Learning.

De Lange, J., Burrill, G., Romberg, T. A., & Van Reeuwijk, M. (1993). *Learning and testing mathematics in context: The case: Data visualization.* Madison, WI: National Center for Research in Mathematical Sciences Education.

Devlin, K. (2012). *Introduction to mathematical thinking.* Petaluma, CA: Devlin.

Dickinson, P., Eade, F., Gough, S., & Hough, S. (2010). Using Realistic Mathematics Education with low to middle attaining pupils in secondary schools. In M. Joubert & P. Andrews (Eds.), *Proceedings of the British Congress for Mathematics Education April 2010.* http://www.bsrlm. org.uk/wp-content/uploads/2016/02/BSRLM-IP-30-1-10.pdf.

Doorman, L. M. (2005). *Modelling motion: From trace graphs to instantaneous change.* Utrecht, The Netherlands: CD-Bèta Press/Freudenthal Institute, Utrecht University.

Doorman, M., Drijvers, P., Dekker, T., Van den Heuvel-Panhuizen, M., De Lange, J., & Wijers, M. (2007). Problem solving as a challenge for mathematics education in The Netherlands. *ZDM Mathematics Education, 39*(5–6), 405–418.

Drijvers, P. (2003). *Learning algebra in a computer algebra environment. Design research on the understanding of the concept of parameter.* Utrecht, The Netherlands: CD-Bèta Press/Freudenthal Institute, Utrecht University.

Drijvers, P. (2015). *Denken over wiskunde, onderwijs en ICT [Thinking about mathematics, education and ICT] [Inaugural lecture].* Utrecht, The Netherlands: Utrecht University.

Drijvers, P., De Haan, D., & Doorman, M. (submitted). *Scaling up mathematical thinking: A case study on the implementation of curriculum reform in the Netherlands.* Manuscript submitted for publication.

Freudenthal, H. (1968). Why to teach mathematics so as to be useful. *Educational Studies in Mathematics, 1,* 3–8.

Freudenthal, H. (1973). *Mathematics as an educational task.* Dordrecht, The Netherlands: Reidel.

Freudenthal, H. (1977). Bastiaan's experiment on Archimedes' principle. *Educational Studies in Mathemtics, 8,* 2–16.

Freudenthal, H. (1991). *Revisiting mathematics education. China lectures.* Dordrecht, The Netherlands: Kluwer Academic Publishers.

Goffree, F. (2002). Wiskundedidactiek in Nederland. Een halve eeuw onderzoek [Mathematics didactics in the Netherlands. A quarter of a century research]. *Nieuw Archief Wiskunde, 5/3*(3), 233–243.

Goffree, F., Van Hoorn, M. M., & Zwaneveld, B. (Eds.). (2000). *Honderd jaar wiskundeonderwijs [One hundred years of mathematics education].* Leusden, The Netherlands: Nederlandse Vereniging van Wiskundeleraren.

Gravemeijer, K. P. E. (1994). *Developing realistic mathematics education*. Utrecht, The Netherlands: CD-Bèta Press/Freudenthal Institute, Utrecht University.

Grossman, R. (1975). Open-ended lessons bring unexpected surprises. *Mathematics Teaching, 71*, 14–15.

Hough, S., Solomon, Y., Dickinson, P., & Gough, S. (2017). *Investigating the impact of a Realistic Mathematics Education approach on achievement and attitudes in Post-16 GCSE resit classes*. Manchester, UK: MMU/The Nuffield Foundation. https://www2.mmu.ac.uk/media/mmuacuk/content/documents/education/Nuffield-report-2017.pdf.

Julie, C., Cooper, P., Daniels, M., Fray, B., Fortune, R., Kasana, Z., et al. (1999). 'Global graphs': A window on the design of learning activities for outcomes-based education. *Pythagoras, 46*(47), 8–20.

La Bastide-van Gemert, S. (2015). *All positive action starts with criticism. Hans Freudenthal and the didactics of mathematics*. Berlin: Springer.

Lewis, H. A. (1994). *'n Ontwikkelingsondersoekstudie na realistiese meetkunde on ondderig in standard ses [A developmental research study on realistic geometry teaching in standard six (grade 8)] (Unpublished master's thesis)*. Bellville, South Africa: University of the Western Cape.

National Center for Research in Mathematical Sciences Education (NCRMSE) & Freudenthal Institute (1997–1998). *Mathematics in context: A connected curriculum for grades 5–8*. Chicago: Encyclopaedia Britannica Educational Corporation.

Noël, G. (1993). La réforme des maths moderne en Belgique [The reform of modern mathematics in Belgium]. *Mathématique et Pédagogie, 91*, 55–73.

Schoenfeld, A. H. (1992). Learning to think mathematically: Problem solving, metacognition, and sense making in mathematics. In D. Grouws (Ed.), *Handbook for research on mathematics teaching and learning: A project of the national council of teachers of mathematics* (pp. 334–370). New York: Macmillan.

Sembiring, R. K., Hadi, S., & Dolk, M. (2008). Reforming mathematics learning in Indonesian classrooms through RME. *ZDM Mathematics Education, 40*, 927–939.

Smid, H. J. (2018 in press). Dutch mathematicians and mathematics education—A problematic relationship. To be published in M. Van den Heuvel-Panhuizen (Ed.), *Reflections from inside on the Netherlands didactic tradition in mathematics education*.

Stacey, K., Almuna, F., Caraballo, M. R., Chesné, J., Garfunkel, S., Gooya, Z., et al. (2015). PISA's influence on thought and action in mathematics education. In K. Stacey & R. Turner (Eds.), *Assessing mathematical literacy—The PISA experience* (pp. 275–306). Cham, Switzerland: Springer.

Streefland, L. (1993). The design of a mathematics course. A theoretical reflection. *Educational Studies in Mathematics, 25*(1–2), 109–135.

Treffers, A. (1978). *Wiskobas doelgericht [Wiskobas goal-directed]*. Utrecht, The Netherlands: IOWO.

Treffers, A. (1987a). *Three dimensions. A model of goal and theory description in mathematics instruction—The Wiskobas project*. Dordrecht, The Netherlands: Reidel Publishing Company.

Treffers, A. (1987b). Integrated column arithmetic according to progressive schematisation. *Educational studies in Mathematics, 18*, 125–145.

Treffers, A. (1991). Meeting innumeracy at primary school. *Educational Studies in Mathematics, 22*, 333–352.

Treffers, A. (2017). Een didactisch tovervierkant. Open opgaven – vrije producties [A didactical magic square. Open problems – free productions]. In M. van Zanten (Ed.), *Rekenen-wiskunde in de 21^e eeuw. Ideeën en achtergronden voor primair onderwijs [Mathematics in the 21st century. Ideas and backgrounds for primary school]*. Utrecht/Enschede, The Netherlands: Utrecht University/SLO.

Treffers, A., & De Moor, E. (1990). *Proeve van een nationaal programma voor het reken-wiskundeonderwijs op de basisschool. Deel 2: Basisvaardigheden en cijferen [Toward a national programme for mathematics education in primary school. Part 2: Basic skills and algorithmic calculation]*. Tilburg, The Netherlands: Zwijsen.

Van den Brink, J., & Streefland, L. (1979). Young children (6-8): Ratio and proportion. *Educational Studies in Mathematics, 10*(4), 403–420.

Van den Heuvel-Panhuizen, M. (1996). *Assessment and realistic mathematics education*. Utrecht, The Netherlands: CD-β Press/Freudenthal Institute, Utrecht University.

Van den Heuvel-Panhuizen, M. (2003). The didactical use of models in Realistic Mathematics Education: An example from a longitudinal trajectory on percentage. *Educational Studies in Mathematics, 54*(1), 9–35.

Van den Heuvel-Panhuizen, M., & Drijvers, P. (2014). Realistic Mathematics Education. In S. Lerman (Ed.), *Encyclopedia of mathematics education* (pp. 521–525). Dordrecht: Springer.

Van der Velden, B. (2000). Between "Bastiaan ou de l'éducation" and "Bastiaan und die Detektive". *ZDM Mathematics Education, 32*(6), 201–202.

Webb, D. C., Van der Kooij, H., & Geist, M. R. (2011). Design research in the Netherlands: Introducing logarithms using Realistic Mathematics Education. *Journal of Mathematics Education at Teachers College, 2*, 47–52.

Chapter 4
The Italian Didactic Tradition

Maria Alessandra Mariotti, Maria G. Bartolini Bussi, Paolo Boero,
Nadia Douek, Bettina Pedemonte and Xu Hua Sun

Abstract Starting with a historic overview highlighting the increasing interest and involvement of the community of mathematicians in educational issues, the chapter outlines some of the crucial features that shaped Italian didactics and, more specifically, the emergence of research studies on mathematics education. Some of these features are related to local conditions, for instance, the high degree of freedom left to the teacher in the design and realization of didactic interventions. The specificity of the Italian case can also be highlighted through a comparison with the reality of other countries. The fruitfulness of this comparison is presented by reporting on collective and personal collaboration experiences between the French and Italian research communities. A final contribution, coming from East Asia, puts the Italian tradition under the lens of a completely new eye, and invites reflection upon historical and institutional aspects of the Italian tradition.

M. A. Mariotti (✉)
University of Siena, Siena, Italy
e-mail: mariotti21@unisi.it

M. G. Bartolini Bussi
University of Modena and Reggio Emilia, Reggio Emilia, Italy
e-mail: mariagiuseppina.bartolini@unimore.it

P. Boero
University of Genoa, Genoa, Italy
e-mail: boero@dima.unige.it

N. Douek
ESPE de Nice, I3DL, CAPEF, Paris, France
e-mail: ndouek@wanadoo.fr

B. Pedemonte
SJSU California, San Jose, USA
e-mail: bettina.pedemonte@gmail.com

X. H. Sun
University of Macau, Macau, China
e-mail: sunxuhua@gmail.com

© The Author(s) 2019
W. Blum et al. (eds.), *European Traditions
in Didactics of Mathematics*, ICME-13 Monographs,
https://doi.org/10.1007/978-3-030-05514-1_4

Keywords Research for innovation · Field of experience · Cognitive unity · Cultural artefacts · Semiotic mediation · Didactic cycle · Semiotic potential of an artefact · Cultural analysis of the content (CAC)

4.1 Introduction

This chapter presents the key features of what we consider to be the Italian didactic tradition.

A short historic overview begins the chapter, showing different voices that illustrate specific aspects of an Italian trend in mathematics education (*didattica della matematica*) from both inside and outside the community of Italian didacticians.

In the next section, a historic overview highlights the increasing interest and involvement of the community of mathematicians in educational issues, in particular, the role played by special personalities in the emergence and the development of mathematics education as a scientific and autonomous discipline.

Paolo Boero and Maria G. Bartolini Bussi then outline crucial features that have shaped Italian didactics and, more specifically, the emergence of research studies in mathematics education; some of these features are related to local conditions, for instance, to the high degree of freedom left to the teacher in the design and the implementation of didactic interventions.

The specificity of the Italian case can also be highlighted through a comparison with the realities of other countries. The fruitfulness of this comparison is presented in a section by Nadia Douek and Bettina Pedemonte, who report and comment on collective and personal collaboration experiences between French and Italian research communities.

A final contribution, coming from East Asia, places the Italian tradition under the lens of a completely new eye. Xu Hua Sun from Macau recently came in contact with the Italian tradition in collaborating with Maria G. Bartolini Bussi on the organization of the 23rd ICMI study.

4.2 Mathematicians and Educational Issues: A Historical Overview

Similar to other European countries such as France, we find great interest and intense engagement from the community of mathematicians in educational issues, starting from Italy's unification and through the 20th century.

The modern state of Italy does not have a long history: The unification of the different regions into one single nation under the monarchy of the Savoy family dates back only to 1861, and only in 1871 was Rome included in the Kingdom of Italy. Immediately after the constitution of the Kingdom of Italy, when the new state began to reorganize the central administration and in particular the educational system, we find eminent mathematicians involved in the Royal committees that were nominated to elaborate curricula for the different schools and engaged in writing textbooks.

In fact, the varied historical and cultural experiences in different parts of the country had created a great variety of textbooks and a substantial dependency on non-Italian books, mainly translations or re-elaborations of French books such as the classic *Elements de Géometrie* by Legendre or of the more recent books by Antoine Amiot, Josef L. F. Bertrand and Josef A. Serret. The flourishing of new textbooks between the end of the 19th century and the beginning of the 20th century testifies to the active role distinguished scholars such as Luigi Cremona and Eugenio Beltrami began to play in supporting the position that mathematics began to assume within the curriculum (Giacardi, 2003). A key issue concerned the recruitment of teachers, since at the time there were no regulations, and people without any specific mathematical competence were allowed to teach. It was only in 1914 that legislation was approved regarding the legal status of mathematics teachers. Many factors determined the active engagement of mathematicians on the social and political front, one of which being the indubitable influence of Felix Klein, which is of interest considering the topic of this chapter (Giacardi, 2012, p. 210). Some of the Italian mathematicians had the opportunity to visit Germany, and they took advantage not only of Klein's scientific training but also of learning about Klein's ideas on reforming secondary and university mathematics teaching. A tangible sign of their appreciation of Klein's project is the early translation into Italian of Klein's Erlangen Program by Gino Fano and Corrado Segre (Giacardi, 2012). Under the influence of the German experience, a wide debate about the structure and the content of the new schools for teachers (*scuole di magistero*) involved many professional mathematicians such as Alessandro Padoa, Gino Loria and Giuseppe Peano. The core of the debate concerned the tension between the pedagogical and disciplinary stances.

An important occasion of international comparison between different experiences in the field of mathematics education was the IV International Congress of Mathematicians which took place in Rome in April 1908. During this congress the International Commission on the Teaching of Mathematics was founded.[1]

However, in spite of what we could define a very promising start, in the 1920s, the Minister of Education, Benedetto Croce, abolished the *scuole di magistero*, so that any kind of teacher training for the secondary school level was cancelled. In the same trend, after the advent of Fascism and with the nomination of Giovanni Gentile as Minister of Education, the role and the influence of mathematics, and generally speaking of scientific disciplines, decreased in the curriculum. The community of mathematicians fought against this trend, defending the cultural value and the fundamental contribution of mathematics and scientific education in general. The most prominent figure in this cultural fight was certainly Federigo Enriques, but other mathematicians such as Guido Castelnuovo, the father of Emma, played a fundamen-

---

[1] Details of the history of this institution can be retrieved at: http://www.icmihistory.unito.it/timeline.php.

tal role, defending the cultural and educational dimension of mathematics, opening specific university courses and producing material for teacher preparation (Marchi & Menghini, 2013). Didactical essays were written discussing original didactical principles, such as dynamic teaching, elaborated by Enriques (1921), and the need for synthesis between intuition and logic argued by Severi (1932, p. 368).

Noteworthy is the cultural project on the elaboration of elementary mathematics content; Enriques (1900, 1912) was the editor of a huge six-volume text entitled *Questioni riguardanti le mathematiche elementari* that collected articles about elementary topics that the authors felt were fundamental for any secondary school mathematics teacher to know.

After the long and traumatic period of the Second World War, the Italian educational system saw the formulation of new curricula for junior high (*scuola media*) and high schools (*licei*). However, in spite of some methodological suggestions inspired by Enriques, the general structure of the curriculum remained unchanged, and a well-settled textbook tradition contributed to crystallise mathematics teaching in schools. Though we can say that Italy was not touched by the New Math 'revolution,' we can find an echo of the international debate in the active discussions that took place in the mathematicians' community. Two particularly interesting occasions were meetings that took place in Frascati (1961–62) leading to a document known as 'Programmi di Frascati', which contained the outline of a curriculum that the Unione Matematica Italiana, the Italian Association of Mathematicians, offered to the Ministry (Linati, 2011). Unfortunately, however, the negative inheritance of Gentile's reform was to last far longer than the new wave, and the Italian upper secondary school had to wait until the academic year 2010–11 to see a reform, both of its structure and of its curricula. However, in spite of the lack or the episodic character of any reform of the curricula or possibly precisely because of this phenomenon, didactics in the Italian school lived a very peculiar development: Never pressed by strict obligations or official curricula, teachers have always had a certain freedom, leaving the possibility of being open to innovations; they could be influenced by new ideas and they could experiment and develop new approaches to math teaching (see the following section for more details).

Though she can certainly be considered unique, it would have been impossible for Emma Castelnuovo to realize the innovative projects that made her so famous if the context of the Italian school had been different.[2] This aspect of the Italian school also explains the influence that, in spite of the immobility of the official curriculum, mathematicians were able to have both in promoting innovations and in developing new research field of mathematics education.

In order to understand the role of the mathematicians' community in this process, it is important to consider the foundation and the role played by the associations of mathematicians. The roles played by individuals become stronger when they are played by representatives of professional associations: This is the case in Italy, where

---

[2]https://www.youtube.com/c9918297-edab-4925-9b95-1642ddbaa6a2; The legacy of Emma Castelnuovo remains in institutions such as the Officina matematica di Emma Castelnuovo (http://www.cencicasalab.it/?q=node/23).

there have been two associations that in different ways have played and still play key roles in promoting mathematics teaching. The first is the Association Mathesis, founded in 1895, and the second is the Unione Matematica Italiana (UMI) founded in 1920 by Vito Volterra as part of the International Mathematical Union. A common characteristic of these associations is the presence of both university and secondary school teachers and the nomination of a permanent committee whose specific role has been to treat educational issues both at the national and at the international levels.

The commitment of the community of mathematicians to educational issues became extremely evident at the beginning of the 1970s, when stable groups of researchers in mathematics education emerged, as will be explained in the following section.

In this respect, we can see a great similarity between the Italian and the French traditions, though for different reasons: The first generation of Italian researchers in mathematics education was made up of professional mathematicians belonging to mathematics departments and often distinguished scholars active in mathematics research fields.

4.3   The Development of an Italian Research Paradigm

In the 1970s in Italy there was a consolidated tradition of cooperation between universities and schools. Famous mathematicians and mathematics teachers were involved in this process: Among the former, it is worthwhile to mention the late Giovanni Prodi, Lucio Lombardo Radice and Francesco Speranza, who engaged in the renewal of the teaching of mathematics in schools with teaching projects, collective documents, and pre- and in-service teacher education activities. Among the latter was Emma Castelnuovo, to whom the ICMI award for outstanding achievements in the practice of mathematics education is dedicated.

Research projects for innovation in the mathematics classroom at different school levels have included mixed groups of academics and teachers at many universities who have developed a cooperative attitude on a basis of parity (Arzarello & Bartolini Bussi, 1998). For more than two decades (1975–1995), the Italian research funding agency, the Italian National Research Council, under the pressure of some mathematicians who were members of the Committee for Mathematical Sciences, funded research groups with the aim of contributing to the renewal of school teaching of mathematics. Research fellowships were allocated to young people engaged in such groups (in particular, at the end of the 1970s).

The simultaneous presence of active school teachers (who later were called teacher-researchers; see Malara & Zan, 2008) and academics allowed the creation of a very effective mixture of practical school needs and an academic tradition: The constraints of the local contexts had to meet international research trends. For instance, the teacher-researchers highlighted the importance of whole-class interaction (beyond the more popular studies on individual problem solving and small-group cooperative learning), the teacher's role as a guide (beyond the more popular focus on learners' processes), long-term processes (beyond the more popular studies on short-term processes) and the manipulation of concrete artefacts (e.g., abacuses, curve drawing devices, and tools for perspective drawing) without overlooking the theoretical aspects of mathematical processes.

During the last four decades, specific conditions of the Italian institutional context of mathematics instruction have demanded and allowed the engagement of university mathematicians in the renewal of school teaching of mathematics through the elaboration and experimentation of long-term educational projects:

- Programs and (more recently) guidelines for curricula have not been strictly prescriptive in comparison with other countries; moreover, teachers and textbook authors feel free to interpret them in rather personal ways. The interpretation often follows a transmissive, traditional paradigm, but innovative experiences, even beyond the borders of the ministry texts, have been proposed, thus preparing further evolution of programs and guidelines for curricula.
- The quality of national programs and (more recently) guidelines for curricula have been rather good: This also depends on the engagement of researchers in mathematics education, mathematics educators and mathematicians in the elaboration of these guidelines (within the Ministry commissions that prepare them). In particular, the Guidelines for Primary and Lower Secondary School Curricula (from 2003 to 2012; MIUR, 2012) were strongly influenced by the previous work of a commission that included mathematicians, statisticians, researchers in mathematics education, teacher-researchers and delegates of the Ministry of Education (see MIUR, SIS & UMI-CIIM, 2001).

This was the context where the present research in didactics of mathematics arose in Italy, with some original features that were later appreciated at the international level.

4.3.1 The Early International Presentations of the Italian Didactics of Mathematics

In the 1980s, when Italian researchers started to take part in international conferences on mathematics education (e.g., ICME and PME), the difference between what was happening in Italy and elsewhere appeared quite evident. In Italy there was not a consolidated tradition of cognitive studies such as those developed in the US and other countries, but there was a tradition of epistemological and historical analysis of content and a growing pressure on active teachers to become an important part of the innovation process.

In 1987, the first session of the Italian National Seminar on Didactics of Mathematics (later dedicated to Giovanni Prodi; see http://www.airdm.org/sem_naz_ric_dm_3.html) took place. In those years, the Italian community started to publish reports on mathematics education research in Italy for the ICME (ICME 9 and ICME 10). They were collective volumes and had a strong effect not only on the external presentation of the Italian traditions, but also on the internal construction of the identity of the Italian community.[3]

This situation was summarized in the 1990s by an invited contribution to ICMI Study 8 (Arzarello & Bartolini Bussi, 1998). The paper summarized a discussion that had been going on in those decades and presented a systemic view, where research for innovation was related to the most popular trends of epistemological analysis and of classroom teaching experiments and to the new trend of laboratory observation of cognitive processes. This choice aimed at pointing out the specific positive features of our action research in the mathematics classroom and answering criticisms coming from other traditions, where the tendency towards more theoretical studies was prevalent.

4.3.2 A Retrospective Analysis 30 Years Later: Some Examples from Classroom Innovation to Theoretical Elaborations

This systemic approach is still fundamental, although, on the one hand, additional components have strongly emerged and are well acknowledged at the international level (e.g., for studies on beliefs and affect issues, see Di Martino & Zan, 2011, and for multimodality in the teaching and learning of mathematics, see Arzarello & Robutti, 2010) and, on the other hand, the landscape has changed with, for instance, the impact of technologies, the attention to students with special needs, the major issue of teacher education, and the awareness of multicultural aspects.

In this section we report on some examples of mathematical education research chosen from those focused upon during specific sessions of the National Seminar and reconstruct their roots. These projects and others (e.g., the ArAL project on the early approach to algebraic thinking; see http://www.progettoaral.it/aral-project/) have been developed by groups of academic researchers and teacher-researchers in many parts of the country.

The field of experience construct

Teaching and learning mathematics in context was an initial aim for a five-year project for elementary school (now referred to as primary school) and a three-year project for lower secondary school planned during 1976–80 and gradually implemented in the 1980s in more than 200 classes (Boero, 1989). The initial roots of these projects were ideas that were circulating both in other countries (e.g., the School Mathematics Project in the UK) and in Italy (e.g., *Mathématiques dans la réalité* or "mathematics in reality"; Castelnuovo & Barra, 1980) together with general educational aims related to education for citizenship. The projects aimed at preparing students to deal with outside-of-school problem situations and horizons and

---

[3]http://www.seminariodidama.unito.it/icme9.php;   http://www.seminariodidama.unito.it/icme10.php.

at establishing links between learning aims and opportunities offered by students' outside-of-school experiences. Gradually, the researchers involved in the projects realized that such didactic and cultural-educational choices needed suitable frameworks in order to establish a relationship between outside-of-school culture(s) and classroom activities. The field of experience construct, proposed by Boero (1989) and then implemented in other studies (see Bartolini Bussi, Boni, Ferri, & Garuti, 1999; Dapueto & Parenti, 1999; Douek, 1999) was one of the answers to that need. A field of experience concerns areas of outside-of-school culture and actual or potential experiences for students, such as in the 'sun shadows' and 'money and purchases' examples, and also geometry that is intended not as school geometry but as geometry that applies to outside-of-school culture, including the geometry of the carpenter example, and not only mathematicians' geometry. A complex relationship can be established between outside-of-school culture, the goals of school teaching and classroom activities through the mutual influence between the three components of a field of experience: teachers' inner contexts, including their knowledge, conceptions, experience and intentions related to the field; the students' inner contexts, including their conceptions and experience; and the external context, particularly the artefacts related to the field of experience, such as material objects and external representations together with customary practices.

Within a given field of experience, classroom activities are conceived as opportunities offered to the teacher to intervene on the students' inner contexts through practices and tools belonging to the external context in order to develop students' everyday concepts into scientific concepts. Specifically, concepts used as tools to deal with typical problems of the field of experience may be further developed as objects belonging to disciplinary fields of experience such as geometry or arithmetic. The field of experience construct was used as a tool to plan and manage long-term classroom activities according to a methodology (for field of experience didactics, see Boero & Douek, 2008) that was essentially based on students' individual or small group productions that become the object of classroom discussions orchestrated by the teacher and are then brought towards progressive phases of institutionalization.

By the end of the 1990s, another need emerged from experimental activities: to develop a cultural framework to deal with cultural diversities in the classroom. Indeed, as far as outside-of-school culture and practices enter the classroom as objects of experimental activities, cultural approaches to outside-of-school realities (e.g., random situations and natural and social phenomena) may be quite diverse and different from how they are dealt with in the classroom. Moreover, within mathematics and between mathematics and other disciplines, knowing, acting and communicating show specificities that must be taken into account in order to develop specific learning activities. As an example, we consider mathematical modelling and proving in mathematics: criteria for the truth of statements and strategies and ways of communicating results are only partially similar, while differences concern relevant aspects, for instance, how models fit the modelled reality versus how the truth of statements derives from axioms and already proved theorems. In order to cope with all these aspects, Habermas' (1998) construct of rationality, which deals with knowing, acting and communicating in discursive practices in general, was adapted to specific discursive practices in mathematical fields of experience and in other fields of experience (Boero & Planas, 2014).

While dealing in our projects with the development of proving skills in Grades 7 and 8, teachers and researchers identified a phenomenon that could result in the choice of suitable tasks to ameliorate students' approach to proving: the cognitive unity of theorems (Boero, Garuti, & Lemut, 2007; Garuti, Boero, Lemut, & Mariotti, 1996), which consists in the possibility (for some theorems) of establishing a bridge between conjecturing and proving. During the conjecturing phase, students produce ideas and develop ways of reasoning that can be resumed and reorganized in the proving phase as components of proving and proof. Other researchers (in particular, Fujita, Jones, & Kunimune, 2010; Leung & Lopez Real, 2002; Pedemonte, 2007), using the theoretical construct of cognitive unity, investigated obstacles and opportunities inherent in the transition from conjecturing and proving in different learning environments.

The theory of semiotic mediation

A pair of projects chaired by one of the authors of this contribution was briefly reported on during PME 22 (Bartolini Bussi, 2009), where a rich list of references is contained. The two projects are complementary and address different school levels. They both have produced papers published in scientific journals and volumes and, in the last decade, they have converged into a larger project on cultural artefacts: semiotic mediation in the mathematics classroom.

The first project concerns a Vygotskian approach to social interaction, presented for the first time in 1991 (in a plenary speech at PME 15). In this project the theoretical construct of mathematical discussion orchestrated by the teacher was elaborated with teacher-researchers (Bartolini Bussi, 1996). A mathematical discussion is

> a polyphony of articulated voices on a mathematical object, that is one of the leitmotifs of the teaching-learning activity. The term voice is used after Wertsch (1991), following Bakhtin, to mean a form of speaking and thinking, which represents the perspective of an individual, i.e., his/her conceptual horizon, his/her intention and his/her view of the world. (Bartolini Bussi, 1996, p. 16)

The teacher utters many different voices, among which the voice of mathematics as it has been developed within the culture where the teaching is being realized. The idea of mathematical discussion orchestrated by the teacher was shared by another research group (see the field of experience didactics mentioned above).

The second project concerns the mathematical laboratory approach centred on the use of concrete instruments. The concrete instruments are, in this case, mathematical machines (e.g., abaci, pair of compasses and curve drawing devices, and perspectographs) taken from the history of mathematics and reconstructed and used for didactical purposes. They are true cultural artefacts (Bartolini Bussi, 2010).

Although the groups of teacher-researchers belonged to different school levels (primary school and secondary school), from a theoretical point of view the two projects were consistent with each other and both referred to two crucial aspects of the Vygotskian tradition: language in social interaction and cultural artefacts in

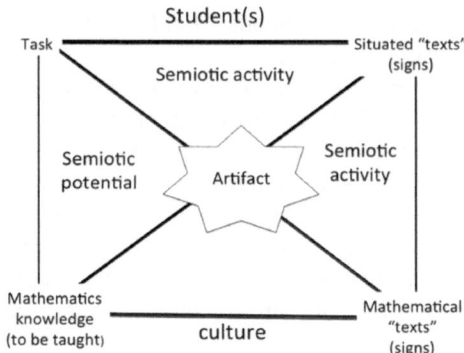

Fig. 4.1 Semiotic mediation

mathematical practices. In the following years, thanks to the contribution of Mariotti (2012), new technologies also entered this joint project, adding information and communication technologies to concrete artefacts and exploiting the effectiveness of mathematical discussion. In this way a comprehensive approach to studying semiotic mediation by artefacts and signs in the mathematics classroom (Bartolini Bussi & Mariotti, 2008; Mariotti, 2009, 2012) was developed. The theory of semiotic mediation using a Vygotskian approach aims to describe and explain the process that starts when students use an artefact to solve a given task and then leads to the students' appropriation of a particular piece of mathematical knowledge. The semiotic potential of an artefact refers to the double semiotic link that may be established between an artefact, the students' personal meanings (emerging from its use when they try to accomplish a task) and the mathematical meanings evoked by its use and recognized as mathematics by an expert. The process of semiotic mediation consists of developing initial situated signs, which evoke the artefact, into mathematical signs referring to mathematical knowledge. In this process, pivot signs, hinting at both the artefact and the mathematical knowledge, appear and are exploited by the teacher. This process is shown in Fig. 4.1. On the left is the triangle of the semiotic potential of the artefact. The teacher plays two different roles in this scheme in the development from artefact signs to mathematical signs: (1) the role of task design (on the left) and (2) the guidance role (on the right). This development is promoted through the iteration of didactical cycles (Fig. 4.2), where different categories of activities take place in the unfolding of the semiotic potential of the artefact: individual or small group activities with the artefact to solve a given task; individual production of signs of different kinds (e.g., drawing, writing or speaking, and gesturing); and collective production of signs, where the individual productions are shared. The collective phases are realized by means of mathematical discussions.

These projects were all originally developed for classroom innovation but later influenced the curricula prepared by the Italian Commission on Mathematical Instruction and adopted by the Ministry of Education; they also were included in handbooks for teachers (e.g., in the Artefatti Intelligenti series: https://www.erickson.it/Ricerca/

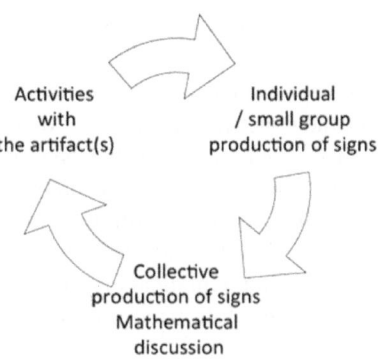

Fig. 4.2 Didactic cycle

Pagine/results.aspx?k=artefatti%20intelligenti). Hence, they proved to be effective at the institutional level.

4.3.3   *The Impact of Italian Research at the International Level*

At the international level, Italian research activities such as those exemplified in the previous sections produced dozens of research papers published in the proceedings of international conferences and in major scientific journals and volumes by international publishers. The two main foci of the Italian contribution were the historical-epistemological dimension and original didactical implementations, which were carried out by mixed groups of academics and teacher-researchers following the steps of well-known personalities of the previous generation (e.g., Emma Castelnuovo). This contribution had the potential to support the dialogue with other dimensions that were more prevalent in other research traditions (e.g., cognitive, sociological and affective). Thus, in spite of the fact that in Italy there has been and still is a structural difficulty related to the discontinuity of financial support and available jobs at the university level, international collaborations have been developed and Italian scholars have reached important positions within various international boards (ICMI, ERME and PME).

Some of the ideas carried out by Italian scholars (e.g., the teacher's role as a guide and a mediator in mathematical classroom processes and the focus on historical-epistemological issues) were not aligned with some widespread assumptions and systems of values shared by other communities who put more emphasis on non-mathematical issues (e.g., sociological and psychological). For example, we may consider the debate between the constructivist positions very popular in US and the historical-cultural trend developed by scholars following the Vygotskian tradition (e.g., Bartolini Bussi, Bertolini, Ramploud, & Sun, 2017; Vianna & Stetsenko, 2006; Waschescio, 1998).

As examples of leading positions that are different from those elaborated by Italian researchers, let us consider the constructs of negotiation of meanings and of socio-mathematical norms (Cobb & Yackel, 1998) that have resulted from a relatively radical vision of constructivism, where the focus is on the learners as comparatively isolated subjects, that has a more or less explicit separation from the historical-cultural context. In those approaches, the very formulation of the teacher's actions is only weakly linked to the specific mathematical task at stake. For instance, Cobb and Yackel (1998), while discussing the solution of a problem in the observed classrooms, state that

> the teachers rarely asked whether anyone had a more sophisticated or more efficient way to solve a problem. However, their reactions to students' solutions frequently functioned as implicit evaluations that enabled students to infer which aspects of their mathematical activity were particularly valued. . . . These implicit judgements made it possible for students to become aware of more developmentally advanced forms of mathematical activity while leaving it to the students to decide whether to take up the intellectual challenge. (p. 169)

This short excerpt indicates that teachers should be cautious to avoid any form of explicit teaching, such as introducing from outside (e.g., from the textbook or a story about a different classroom) a more effective strategy, even though it is part of the historical development of mathematics.

Drawing on this literature, in the Italian tradition we have stressed the necessity for a clear focus on the specificity of the piece of mathematics knowledge at play, on the particular task and the artefacts proposed to the students, on the cultural heritage to be exploited in the classroom, and on the related cultural roots of mathematics teaching and learning.

Yet our background choices seem to be consistent with the systems of values and cultural traditions developed in other large regions of the global world that have appeared only recently in the forum of mathematics education (e.g., East Asian regions and developing countries). It is worthwhile to mention the very recent paper by Bartolini Bussi, Bertolini, Ramploud, and Sun (2017), where some assumptions of the theory of semiotic mediation are integrated into the Chinese approach to lesson study and exploited to carry out teaching experiments in Italian classrooms. Chinese lesson study is linked to but less known than the Japanese lesson study (Huang et al. 2017). Both are robust methodologies aimed at improving classroom practice by means of a collective design of a lesson, a collective observation of the lesson implemented by a teacher of the design group, and a collective analysis and re-design of the lesson. In the Chinese model, the focus is on the content and on the instructional coherence of the lesson (Mok, 2013), which highlights the implicit or explicit interrelation of all mathematical components of the lesson. Hence from the epistemological perspective, the focus is consistent with the theory of semiotic mediation, where the mathematics knowledge is highlighted. In spite of this episte-mological consistency, however, differences appear when the didactical aspects of classroom organization are considered; for instance, experiments carried out in Italy introduced teaching methodologies developed in the West (e.g., cooperative learning for small group work and mathematical discussion) and observation of the learning process highlighting the development from artefact signs (including drawing, verbal utterances and gestures) to mathematical signs. This was possible thanks to a careful cultural transposition (Mellone, Ramploud, Di Paola, & Martignone, 2017) 'where the different cultural backgrounds generate possibilities of meaning and of mathematics education perspectives, which in turn organize the contexts and school mathematics practices in different ways' (Mellone & Ramploud, 2015, p. 571). This difference appears not only in the choice of artefacts (in the above case, cultural artefacts) but also in the choice of the analytical tools, and related theoretical elaborations may be related to the cultural context (Boero, 2016).

The issue of the cultural contexts is coming to the foreground in the international community: It was chosen as a specific focus in the Discussion Document of the ICMI Study 23,[4] it was addressed at a plenary forum in CERME 9 (Jaworski, Bartolini Bussi, Prediger, & Novinska, 2015) and it was focused on by Barton (2017) in his plenary address, "Mathematics, education and culture: a contemporary moral imperative" in ICME 13.[5] It was acknowledged also by Clarke (2017) during his plenary address at PME 41 in Singapore:

> Within any educational system, the possibilities for experimentation and innovation are limited by more than just methodological and ethical considerations: They are limited by our capacity to conceive possible alternatives. They are also limited by our assumptions regarding acceptable practice. These assumptions are the result of a local history of educational practice, in which every development was a response to emergent local need and reflective of changing local values. Well-entrenched practices sublimate this history of development. In the school system(s) of any country, the resultant amalgam of tradition and recent innovation is deeply reflective of assumptions that do more than mirror the encompassing culture: They embody and constitute it. (Clarke, 2017, p. 21)

The same author (Clarke, personal communication) observed that in several languages (such as Japanese, Dutch, Russian, Chinese and Polish[6]) some communities have acknowledged the interdependence of instruction and learning by encompassing both activities as one process and, most significantly, labelled with a single word, whilst in English, we seem compelled to dichotomise classroom practice into teaching and learning.

4.3.4   Mathematics Teacher Education and Professional Development in Italy

In spite of the rich, decades-long development of the Italian research, the optimistic illusion of a gradual expansion to schools throughout the country was overcome

---

[4] http://www.mathunion.org/fileadmin/ICMI/docs/ICMIStudy23_DD.pdf.

[5] https://lecture2go.uni-hamburg.de/l2go/-/get/v/19757.

[6] Japanese: *gakushushido*; Dutch: *leren*; Russian: *obuchenie;* Chinese: *xue;* Polish: *uczyć*. Also in ancient Italian, *imparare* had both meanings.

and the problematic issue of mathematics teacher education and development came into the foreground. The story of teacher education and development in Italy is recent. The model for pre-primary and primary school dates back to 1998, with the creation of a university program (with a 4- to 5-year master's degree) that has a balance between university courses, workshops and practicums (now 600 h). In nearly 20 years, a good balance between the different components has been reached thanks to the strong coordination of a national conference of all the deans of the university programs and to the realization of specific scientific national seminars. In particular, for mathematics there are, in all the universities where the program is implemented (about 30 throughout the country), 22 European credits[7] for mathematics. The story of secondary school teacher education is more discontinuous. After a start in 1999, with nine cycles of two years each (specialization after a master's degree in mathematics or in other scientific subjects), the process was interrupted and has only recently started again. The most problematic issue seems to be the real collaboration between the general educators and the subject specialists within the discussion about a correct balance between 'pedagogical' issues and specific content. The situation is in progress now.

In recent years, the development of pre-service teacher education at the university level has stimulated Italian research on this subject. In particular, we think that teachers' cultural awareness of the reform processes happening in mathematics should be developed in countries (such as Italy) in which teachers' autonomy is great in implementing national programs and guidelines for curricula. These considerations suggested the elaboration (Boero & Guala, 2008) and implementation (Guala & Boero, 2017) of the construct of cultural (epistemological, historical and anthropological) analysis of the content (CAC). CAC is a competence that should be important to develop in mathematics education courses, as a way of looking at the mathematical content presented.

Pre-service teacher education is only a part of the professional development of teachers: In-service teachers are required to continue their education, thus there is a large number of in-service teachers in development programs. In the past, in-service education (i.e., teacher development) was usually left to individual choices or the local choices of some private institutions, with a few but important exceptions. As an example, we can quote the national program for in-service primary teacher education developed after the reform of national programs for primary schools. The new programs were issued in 1985; in the following years primary teachers were involved in compulsory teacher education initiatives promoted by the regional school supporting institutions, with a strong engagement in many regions with university researchers in mathematics education and teacher-researchers. In recent years, the Ministry of Education has promoted a set of initiatives (MIUR, 2010) aimed at making the image of mathematics and sciences more attractive for pre-university school students in order to try to counter the decrease in the number of students in mathematics and sciences at the university level. Among the initiatives of the Ministry, the so-called laboratory was an original kind of in-service teacher education

---

[7]https://ec.europa.eu/education/sites/education/files/ects-users-guide_en.pdf.

activity: University researchers (in particular, researchers in mathematics or science education) joined with teacher-researchers and teachers to plan innovative classroom activities on important content in various disciplines. Planning was followed by classroom implementation of the activities designed by the teachers. A team analysis of what happened in the classroom was produced and reports on those innovative experiences were disseminated.

Since the school reform in 2015 (the so-called *la buona scuola*, 'the good school'; see https://labuonascuola.gov.it), in-service teacher development has become compulsory, systematic and funded for the first time in Italy's history.

4.4   Collective and Personal Experiences of Collaboration Between French and Italian Researchers

In this section, a rich collective experience will be presented from the perspective of two researchers: Nadia Doueck and Bettina Pedemonte. These contributions reflect aspects and exemplar modes of collaboration between French and Italian researchers.

4.4.1   The Séminaire Franco Italien de Didactique de l'Algèbre

The Séminaire Franco Italien de Didactique de l'Algèbre (SFIDA) was founded in 1993 by Ferdinando Arzarello, Giampaolo Chiappini and Jean-Philippe Drouhard and was held until 2012.[8] It was a full-day seminar that took place alternatively in Nice and Genoa or Torino twice a year. During the seven hours of the seminar, six presentations were programmed around a main theme. The allotted time was sufficient to develop deep discussions, in French and Italian, and if needed, in English. A French-Italian vocabulary was displayed to help understand terms that were quite different in the two languages (such as *élèves-alunni*). Abstracts were shared before the meeting to foster interactions. Work documents have been collected in three volumes up to the present.

The stable group of participants (including the promoters and Assude, Bagni, Bazzini, Boero, Douek, Malara, Maurel and Sackur) did not form a research team, but collaborations flourished now and then between the participants and with others who would occasionally attend the seminar. Researchers whose work related to the theme of the meeting were often invited to attend the seminar. Invited researchers included Duval, Hanna, Radford, Tsamir and Vergnaud.

---

[8] The reader will find the presentation of SFIDA, the French-Italian seminar of algebra's didactics, written by the late Jean-Philippe Drouhard in 2012, in the introduction of a special issue on Algebra published by RDM: (REF).

Openness was the main and constant characteristic of our meetings, in terms of the choices of themes, the variety of theoretical frameworks and the arguments used within the debates: Among theoretical references that framed presentations and backed the arguments exchanged and discussed were conceptual fields, experience field didactics, local knowledge and triple approach, embodiment, semiotic registers, semiotics, theory of didactic situations, anthropological theory of didactics and rational behaviour.

There was also openness in the wide extension to connected areas such as epistemology and history of mathematics, philosophy, and cognitive sciences.

Debates were not bound to remain in one particular framework nor at a particular theoretical level. This openness and variety stimulated creativity, favoured advancements and often led to more precise productions. More than 20 PME research reports were inspired and supported by the work done in SFIDA.

SFIDA has also had an important impact on international collaborations: It not only prepared some of the participants to take on responsibilities within PME and ICMI, but it also had a special role in the foundation of ERME. J. Philippe Drouhard and Paolo Boero were among the founders of ERME and were the first two presidents. In fact, the type of collaboration used in SFIDA deeply inspired the organization of CERME's working groups.

4.4.2   Encounter with Various Intellectual Traditions and Methods in French-Italian Ph.D. Projects

A personal testimony by Nadia Douek

I attended various seminars in France and in Italy, specifically some of the regular meetings of Genoa's research group. The span of contents and the styles of interaction I was exposed to were quite different but all very stimulating. It was in this rich frame that my Ph.D. project started, under the supervision of Gerard Vergnaud (Université Paris V Sorbonne) and Paolo Boero (Università di Genova). Grounded in the French and Italian education traditions, my thesis developed through experiencing different types of relations between theory and class activity and between a priori/a posteriori analysis and the flow of class experiments. I learned to combine theoretical clarifications and quick adaptations (improvising!) in classroom contexts in close relation to the teacher's steps. It also brought me to combine references from the French didactics (works of Vergnaud and influences from Douady and Duval) with Italian didactics, in particular Boero's construct of experience fields.

Vygotsky's *Thought and Language*, a strong reference for several Italian colleagues, and for Vergnaud as well, was also crucial to my intellectual development. Philosophers such as Deuleuze and epistemologists such as Feyerabend that I discovered in French seminars influenced my ways of studying.

Class experiments were central to my thesis work. They took place in Italian classes within the long-term projects developed by the Genoa research team led by

Boero. I had the opportunity to explore and co-develop experimental situations in relation to my participation in the Italian seminars. I was introduced to the Italian colleagues' way of intertwining theoretical with very pragmatic questioning about ongoing class activity developed by the teachers within their research for innovation (see Sect. 3). In this method, research problems are scrutinized by the team under the lens of teaching and learning difficulties in classes. While the theoretical frame is a tool to understand, predict, design didactical settings and so forth, it is questioned through long-term and repeated experimentations. Theoretical components are developed on the basis of the analysis of experienced situations. The support of the teachers—who are involved as researchers in the team—was crucial in my scientific development.

I do not know to what extent my scientific development depended on national scientific trends or on scientific personalities. To characterise some of the methodological or cultural differences between my two supervisors, I can say that working with Vergnaud allowed me to explore the limits of various theoretical frames and to elaborate fruitful ideas even when these were not mature. Working with Boero helped me to mature these ideas into more organized ones and to develop a consciousness of their theoretical limits or validity.

Experimental work in the classroom fostered my creativity and improved my attention to pupils' activity, freed from a priori expectations. Most importantly, it made me understand the fruitfulness of slowing down the pace of class activity and deepening it, but also it allowed me to share intellectual ambitions with the teachers and with the pupils. This is generally difficult to achieve in French classes where curricula are more closely followed with stricter time organization through the year and with a stricter epistemological perspective on mathematics.

The research for innovation methodology was combined with Vergnaud's influence, and the methods of organizing class activity, as practiced within Boero's team, had a very harmonious resonance with my philosophical interests nourished by the Parisian seminars I attended. In particular, they had a great effect on my creativity and conception of knowledge as dynamic, always intertwined with activity and swinging between focusing and questioning.

The experience with Boero's research group also affected my vision of teacher training. Research for innovation implied a specific human organization and composition of the team. Today I can compare it with Engeström's elaboration on expansive learning: Teachers' learning and professional development as research team members relied on collaborative relations, tending to be horizontal. The team members were teachers and university researchers involved in matters of various disciplines. They were all productive. Their oral interventions, transcriptions of class discussions, analysis of students' productions, critical analysis of didactical settings, cooperation to transform them or produce new ones and other activities modulated the global evolution of the group's activity. Their discussions and reflections had an impact on productions involving various questions from very practical to highly theoretical or philosophical ones.

Following their activities and collaborating with Italian researchers until the present, I can still see the richness and openness I enjoyed during my studies. However, I think I can also see that it has not appeared to spread. Therefore, a question arises: Is this methodology, reflecting a leading motive in Italian research, still operating as strongly and productively as in the past?

Two factors may have weakened this trend: a favouring of theoretical development per se (less contrasted by long-term class experiments and by multilevel and multi-spectrum discussions) and the need to systematize class activities and designs under the pressure of seeking visible, stabilized utilitarian products.

If this is the case, I fear that it might restrain the evolution of didactic research into more narrow paths that are more tightly managed and that it would weaken the collaboration with teacher-researchers, especially if the spirit of the research is to stabilize results in too firm a manner. I do however remain confident in its powerful potential at both the scientific level and the social and human levels.

A personal testimony by Bettina Pedemonte

In 1997, I obtained my BS degree in Mathematics in Italy at Genoa University. My supervisor, Paolo Boero, presented the results of my study at the 21st PME Conference and invited me to participate to the congress. It was my first experience as a participant of a didactical conference and I was immediately fascinated by this new world. I decided to pursue my studies to earn a Ph.D. in mathematics education. At that time, there was not a specific program in this field in Italy. So I went to France, and after earning my degree, a *Diplôme d'etude approfondi* (DEA), I continued my studies within a joint agreement between the universities of Grenoble, France, and Genoa, Italy. I developed my Ph.D. project under the co-supervision of Nicolas Balacheff from the National Centre for Scientific Research and Maria A. Mariotti from the University of Pisa. I defended my Ph.D. dissertation in June 2002, obtaining two titles: Doctor of Mathematics Education from Grenoble University and Doctor of Mathematics from Genoa University.

The main purpose of my Ph.D. project was to study the relationship between argumentation and proof in mathematics from a cognitive point of view and to analyse how students can be supported in understanding and constructing mathematical proofs.

The hypothesis of my research was that there would usually be continuity, called cognitive unity, between these two processes (Boero, Garuti, & Mariotti, 1996; Garuti, Boero, Lemut, & Mariotti, 1996). The theoretical framework of reference, however, did not provide the tools to validate this hypothesis. I made a scientific contribution to this field by constructing a specific frame for mathematical argumentation with the aim of comparing it with proof. On the basis of contemporary linguistic theories, I designed a theoretical framework that allowed me to 'model' a mathematical argumentation, to compare it with the proof and to describe the cognitive processes involved. This cognitive analysis was based on a methodological tool, Toulmin's (1958) model integrated with the 'conception' model (Balacheff, 2013). This tool and the results obtained using it in different mathematical fields, geometry and algebra, were the innovative aspect of this study.

During the work on my Ph.D. thesis, I learned to study, compare and integrate educational studies that were deeply different from each other from methodological, cultural and scientific points of view. I learned to distinguish and select materials from different cultures and merge them perfectly. My two supervisors were very helpful during this stage. Nicolas Balacheff taught me to always have a solid basis for expressing my opinion and point of view. He taught me the importance of constructing a theory to support my ideas. Maria A. Mariotti helped me to order and organize my thoughts. When I was lost and felt I had too many ideas in my mind, I would talk with her and she would guide me and help me find my focus based on the purpose of my thesis. They both helped me to gain confidence and they always supported me, pushing me to be creative and innovative.

The interconnection between Italian and French cultures was essential for my Ph.D. project. The content of my thesis stems from an Italian notion: that of cognitive unity between argumentation and proof. However, it was in France that I developed the methodological competences to set this notion inside a theoretical framework. The creative and innovative aspects of my thesis probably come from Italian culture, but without the structural frame provided by the French methodology, the results of my work would probably not be so impressive and well accepted by the educational community.

During my Ph.D. work, I designed and implemented experimentations in French and Italian classes. What I observed in these classes not only confirmed my research hypothesis but also provided new insights and clues for future perspectives.

The three geometry problems I used in my experiment during my Ph.D. were solved differently by Italian and French students. In the eighth and ninth grade curricula, geometry proofs in France focus on transformations (translation, rotation and symmetry), while in Italy they are oriented toward the congruence theorems for triangles. Furthermore, the notion of proof for Italian and French students is not completely the same at these school levels. In Italy, when asked for a proof, students know that the proof needs to be deductive (*dimostrazione*). In France, the justification of the conjecture is also accepted when it is not deductive (*preuve* instead of *démonstration*).

Aside from these important differences, the results of my experiments were very similar. In both countries, I observed that sometimes students were unable to construct proofs because they were not able to transform the structure of their argumentation into a deductive structure (for example, some students often construct abductive proofs if they have previously produced abductive argumentations). This aspect was observed not only in France but also in Italy, where students were aware that deductive proofs were expected. Italian students were more reluctant to write proofs that were not deductive. In some cases, they preferred not to write anything at all. Instead, in France, although deductive proofs were explicitly required by the teacher, when students were not able to construct a deductive proof, they handed in informal justifications.

Despite the different curricula and the different conceptions of proof students had in France and in Italy, in both experiments I observed how hard it was for students to construct a deductive poof when they had previously constructed an abductive argumentation. This strong result, confirmed in two different cultures, pushed me to broaden my research. I decided to investigate whether these results could be generalized not only to other mathematical domains but also to other cultural environments (Martinez & Pedemonte, 2014; Pedemonte & Buckbinder, 2011; Pedemonte & Reid, 2010) in my studies concerning students' conceptions in the construction of proofs (Pedemonte & Balacheff, 2016). This research, bridging results from the sciences of language, epistemology and mathematics education, has been going on for more than 10 years. I believe that without my intercultural experience between Italy and France this research would not have been developed to this extent.

During my Ph.D. work in France, I was constantly in contact with other students. I had many discussions with them. We regularly had meetings where we compared our results, helped each other to find appropriate methods to construct our experiments and anticipated possible critical comments from external researchers. It was in this environment that I learned to accept negative feedback as constructive for my research, learning to become a strict reviewer of my own research.

I was very lucky to have two supervisors, but I needed to convince both of them about my new ideas. This taught me to strongly justify my research before sharing it with others. In a broader sense, my research needed to be accepted by two different research communities—Italian and French—with different research methodologies and theoretical frames. This taught me to always consider multiple points of view, to anticipate possible feedback and to integrate research characteristics (theoretical frameworks, methods and experimental research) from different cultures. Even now that I am living in the US, I have realized how important that experience was for my professional career. It would probably be harder to integrate myself into this new community if I had not learned in France how to manage this kind of situation.

Even if this experience was probably one of my best in my entire life, it was not easy to take the decision to move to France to earn my Ph.D. Beyond the difficulties coming from my poor French language skills, the most difficult aspect was to interconnect two different cultures, the Italian inside and the French outside. During the first six months, I refused the new culture: I did not like the French language because it was a real obstacle to me and I did not like the French bureaucracy, which I thought was even worse than the Italian. Also, people appeared strange to me even if they were nice. However, little by little this perception of the French world changed in me and this new culture became part of me, until I recognized both cultures inside me. I finally understood that this experience not only gave me a professional career but also a more flexible mind and an open view on what research consists of.

4.5 The Italian Tradition From a Chinese Cultural Perspective

A personal testimony by Xu Hua Sun

It is not easy for me to clarify what the Italian tradition is due to my limited knowledge. My presentation will comment on some impressive aspects of the Italian tradition from my personal point of view and from a Chinese cultural perspective. My considerations are based on my experience in collaborating with Mariolina Bartolini Bussi on the organization of the ICMI Study 23 (Bartolini Bussi & Sun, 2018; Sun, Kaur, & Novotna, 2015) and aim to foster a reflection on institutional and historical aspects of the Italian tradition.

4.5.1 On the Historical Aspects of the Italian Tradition

There is a long history of relationships between Italian and Chinese people. Marco Polo is perhaps the most famous of Italian visitors. When he visited ancient China in the 13th century, he was not the first European to reach China, but he was the first to leave a detailed chronicle of his experience, and his book opened communication between the East and West. His *Book of the Marvels of the World*, also known as *The Travels of Marco Polo*, c. 1300, was a book that described to Europeans the wealth and great size of China.

Less known to the majority, but well known to mathematicians, is the Italian Matteo Ricci, who in the 17th century collaborated with Xu Guangqi on translating the first six books of Euclid's *Elements* (Volumes 1–6; Chinese title: 幾何原本) that were published and printed in 1607.

This translation offered the Chinese people a first look into the Western tradition of logical deduction. Euclid starts with a list of definitions and postulates in order to reach some clearly stated theorems, providing the details of the necessary steps explicitly. This did not happen in ancient China as '[the Chinese] propose all kinds of propositions but without proof', as Ricci stated (from Dai & Cheung, 2015, p. 13). Most Chinese traditional mathematics concerned arithmetic and algebra that was not supported by deductive reasoning. In the ancient mathematical texts, a typical situation is as follows:

'Very often the presentation starts with a *particular* problem (*wen* in Chinese) being stated in words. Then an answer (*da* in Chinese) is given immediately after that, and from time to time, when it is necessary, the technique or algorithm (*shu* in Chinese) will be outlined' (Dai & Cheung, 2015, p. 13).

Chinese scholars (e.g., Guo, 2010) credit Matteo Ricci as having started China's enlightenment. Deductive reasoning, introduced by the translation of Euclid, had important influences on Chinese mathematics (Sun, 2016a, b). It is considered one of the most important events in Chinese mathematics history, changing Chinese mathematics teaching. The concepts of axioms, theorems, assumptions and proving appeared in the field of mathematics education for the first time. The Chinese terms for points, segments, straight lines, parallel lines, angles, triangles and quadrilaterals were created for the first time and have remained until the present, afterwards having spread to Japan and other countries.

4.5.2 On the Characteristic Aspects of the Italian Tradition

The impact of the Italian geometry culture became evident to me in 2011 when I visited the Laboratory of Mathematical Machines in Modena. The rich collection of more than 200 copies of ancient mathematical instruments stored and displayed in a large room in the Department of Mathematics was impressive to me. Most of these instruments were geometrical instruments (e.g., curve drawing devices and perspectographs) and were evidence of an advanced technological development together with a geometrical knowledge that did not seemed comparable with that of China in the same period. The cultural value of geometry, and in particular its relationship with artistic productions of famous painters such as Leonardo da Vinci, clarified to me the strength of geometry in the Italian didactic tradition and gave me a reason to understand the weakness of a lack of thoughtfulness in the Chinese tradition. This impression was reinforced by some talks that Mariolina Bartolini Bussi and Ferdinando Arzarello gave during the ICMI Study 23 in Macau and by the passionate discussions that followed.

In particular, I was very interested in the didactical setting of the mathematics laboratory as described in the curricula proposed by the Italian Mathematics Association.

Mathematics laboratory activity involves people, structures, ideas and a Renaissance workshop in which the apprentices learn by doing, seeing, imitating and communicating with each other, namely, practicing. In these activities, the construction of meanings is strictly bound both to the use of tools and to the interactions between people working together.

The experimentation-oriented approach that is promoted in the mathematical laboratory setting is different from exercise-oriented learning processes, which are dominated by the manipulation of symbols and a general lack of interest in the real world and are so common in Chinese school practice; at the same time, the experimentation-oriented learning process supported geometric perceptual intuition, for which concrete manipulation seems so fundamental (Zhang, 2016).

A critical comparison of Italian and Chinese culture (Bartolini Bussi & Sun, 2018) highlighted another key element: While the use of the number line can be considered a common feature and a familiar teaching aid in Italian classes throughout the different school levels, Chinese traditional practice nearly ignores it. In fact, the Chinese and Italian traditions are based on different ways of representing numbers. On one hand, representing a number on the number line can be traced back to the geometric roots of Euclid's *Elements* and can embody its geometric continuum as cultural heritage (Bartolini Bussi, 2015). On the other hand, in Chinese culture, numbers are related to the old practice of the abacus and to the verbalization of procedures based on numbers' names embedding a positional representation (Sun, 2015).

4.5.3   On the Institutional Aspects of the Italian Tradition

I had the opportunity to take part in the early implementation of a model of mathematics teacher development in Italian schools inspired by Chinese lesson study (Bartolini Bussi, Bertolini, Ramploud, & Sun, 2017) already mentioned above. This cooperation allowed me to understand many differences between the Italian and Chinese school systems from an institutional perspective.

First, the Italian school system is completely inclusive, whilst students with special needs in China attend special schools. Though it could be easier to plan lessons for homogeneous groups of students, in Italian classrooms, activities must be designed in order to have a positive effect on all students, including low achievers. This requires great care to design lessons for students with very different cognitive levels; nevertheless, the possibility for low achievers to be supported is important for an equitable society from a long-term perspective.

Second, the lesson duration in mainland China is strictly 45 min, without any flexibility, whilst in Italy teachers can adjust the activity a bit if there is not enough time.

Third, attention is paid by Italian teachers to long-term processes because of the autonomy given to schools to plan the whole year of activity, whilst in China, control is strictly exerted by the Ministry of Education. In general, an Italian teacher may have more freedom than a Chinese teacher to introduce innovative activities in order to fulfil the given goals, as time control is not exerted from the outside.

The above observations hint to institutional differences. We must acknowledge that in all cases institutional differences draw on cultural differences, as Italy is in the West, whilst China is under the influence of the so-called Confucian heritage culture (Leung, Graf, & Lopez-Real, 2006, 2015). This dialogue between Italian and Chinese culture is certainly partial and limited, though it is part of a more general dialogue fostered by ICMI in the last decade, as mentioned above (see Sect. 3.3). We hope such dialogue can continue and be deepened, allowing cultures to enrich each other.

References

Arzarello, F., & Bartolini Bussi, M. G. (1998). Italian trends in research in mathematics education: A national case study in the international perspective. In J. Kilpatrick & A. Sierpinska (Eds.), *Mathematics education as a research domain: A search for identity* (Vol. 2, pp. 243–262). Boston: Kluwer.

Arzarello, F., & Robutti, O. (2010). Multimodality in multi-representational environments. *ZDM—The International Journal on Mathematics Education, 42*(7), 715–731.

Balacheff, N. (2013). cK¢, a model to reason on learners' conceptions. In M. Martinez & A. Castro Superfine (Eds.), *Proceedings of the 35th Annual Meeting of the North American Chapter of the International Group for the Psychology of Mathematics Education*. Chicago, IL: University of Illinois at Chicago.

Bartolini Bussi, M. G. (1996). Mathematical discussion and perspective drawing in primary school. *Educational Studies in Mathematics, 31,* 11–41.

Bartolini Bussi, M. G. (2009). In search for theories: Polyphony, polysemy and semiotic mediation in the mathematics classroom. In M. Tzekaki, M. Kaldrimidou, & H. Sakonidis (Eds.), *Proceedings of the 33rd Conference of the International Group for the Psychology of Mathematics Education* (Vol. 2, pp. 121–128). Thessaloniki: Aristotle University of Thessaloniki.

Bartolini Bussi, M. G. (2010). Historical artefacts, semiotic mediation and teaching proof. In G. Hanna, H. N. Jahnke, & H. Pulte (Eds.), *Explanation and proof in mathematics: Philosophical and educational perspectives* (pp. 151–168). Berlin: Springer.

Bartolini Bussi, M. G. (2015). The number line: A "western" teaching aid. In X. Sun, B. Kaur, & J. Novotná (Eds.), *Proceedings of the 23rd ICMI Study 'Primary Mathematics Study on Whole Numbers'* (pp. 298–306). Macao, China: University of Macao. http://www.umac.mo/fed/ICMI23/proceedings.html. Accessed 20 Feb 2016.

Bartolini Bussi, M. G., Bertolini, C., Ramploud, A., & Sun, X. (2017). Cultural transposition of Chinese lesson study to Italy: An exploratory study on fractions in a fourth-grade classroom. *International Journal for Lesson and Learning Studies, 6*(4), 380–395. https://doi.org/10.1108/IJLLS-12-2016-0057.

Bartolini Bussi, M. G., Boni, M., Ferri, F., & Garuti, R. (1999). Early approach to theoretical thinking: Gears in primary school. *Educational Studies in Mathematics, 39,* 67–87.

Bartolini Bussi, M. G., & Mariotti, M. A. (2008). Semiotic mediation in the mathematics classroom. Artefacts and signs after a Vygotskian perspective. In L. English & D. Kirshner (Eds.), *Handbook of international research in mathematics education* (pp. 746–783). New York: Routledge.

Bartolini Bussi, M. G., & Sun, X. H. (Eds.). (2018). *Building the foundation: Whole numbers in the primary grades*. New York: Springer Press. https://www.springer.com/gp/book/9783319635545.

Barton, B. (2017). Mathematics, education, and culture: A contemporary moral imperative. In G. Kaiser (Ed.), *Proceedings of the 13th International Congress on Mathematical Education: ICME-13* (pp. 35–45). https://www.springer.com/gb/book/9783319625966.

Boero, P. (1989). Mathematics literacy for all. *Proceedings of PME-XIII, 1,* 62–76.

Boero, P. (2016). Some reflections on ecology of didactic research and theories: The case of France and Italy. In B. R. Hodgson, A. Kuzniak, & J.-B. Lagrange (Eds.), *The didactics of mathematics: Approaches and issues* (pp. 26–30). Switzerland: Springer International Publishing.

Boero, P., & Douek, N. (2008). La didactique des domaines d'expérience. *Carrefours de l'Education, 26,* 103–119.

Boero, P., Garuti, R., & Lemut, E. (2007). Approaching theorems in grade VIII: Some mental processes underlying producing and proving conjectures, and conditions suitable to enhance them. In P. Boero (Ed.), *Theorems in school: From history, epistemology and cognition to classroom practice* (pp. 249–264). Rotterdam, The Netherlands: Sense Publishers.

Boero, P., Garuti, R., & Mariotti, M. A. (1996). Some dynamic mental processes underlying producing and proving conjectures. In H. L. Chick & J. L. Vincent (Eds.), *Proceeding of the 20th Conference of the International Group for the Psychology of Mathematics Education* (Vol. 2, pp. 121–128). Valencia, Spain: PME.

Boero, P., & Guala, E. (2008). Development of mathematical knowledge and beliefs of teachers: The role of cultural analysis of the content to be taught. In P. Sullivan & T. Wood (Eds.), *The international handbook of mathematics teacher education* (Vol. 1, pp. 223–244). Rotterdam: Sense Publishers.

Boero, P., & Planas, N. (2014). Habermas' construct of rational behavior in mathematics education: New advances and research questions. In P. Liljedahl, C. Nicol, S. Oesterle, & D. Allan (Eds.), *Proceedings of the 38th Conference of the International Group for the Psychology of Mathematics Education* (Vol. 1, pp. 228–235). Vancouver, Canada: PME & UBC.

Castelnuovo, E., & Barra, M. (1980). *Mathématiques dans la réalité*. Paris: Nathan.

Clarke, D. (2017). Using cross-cultural comparison to interrogate the logic of classroom research in mathematics education. In B. Kaur et al. (Eds.), *Proceedings of the 41st Conference of the*

International Group for the Psychology of Mathematics Education (Vol. 1, pp. 13–28). Singapore: PME.

Cobb, P., & Yackel, E. (1998). A constructivist perspective on the culture of the mathematics classroom. In F. Seeger, J. Voigt, & U. Waschescio (Eds.), *The culture of the mathematics classroom* (pp. 158–190). Cambridge: Cambridge University Press.

Dai, Q., & Cheung, K. L. (2015). The wisdom of traditional mathematical teaching in China. In L. Fan, N. Y. Wong, J. Cai, S. Li (Eds.), *How Chinese teach mathematics* (pp. 3–42). Perspectives from insiders Singapore: World Scientific.

Dapueto, C., & Parenti, L. (1999). Contributions and obstacles of contexts in the development of mathematical knowledge. *Educational Studies in Mathematics, 39,* 1–21.

Di Martino, P., & Zan, R. (2011). Attitude towards mathematics: A bridge between beliefs and emotions. *ZDM—The International Journal on Mathematics Education, 43*(4), 471–483.

Douek, N. (1999). Argumentation and conceptualization in context. *Educational Studies in Mathematics, 39,* 89–110.

Enriques, F. (Ed.) (1900). *Questioni riguardanti la Geometria elementare.* Zanichelli, Bologna.

Enriques, F. (Ed.) (1912). *Questioni riguardanti le Matematiche elementari.* Zanichelli, Bologna.

Enriques, F. (1921). Insegnamento dinamico. *Periodico di Matematiche, 4*(1), 6–16.

Fujita, T., Jones, K., & Kunimune, S. (2010). Students' geometrical constructions and proving activities: A case of cognitive unity? In M. M. F. Pinto & T. F. Kawasaki (Eds.), *Proceedings of the 34th Conference of the International Group for the Psychology of Mathematics Education* (Vol. 3, pp. 9–16). Belo Horizonte, Brazil: PME.

Garuti, R., Boero, P., Lemut, E., & Mariotti, M. A. (1996). Challenging the traditional school approach to theorems: A hypothesis about the cognitive unity of theorems. In H. L. Chick & J. L. Vincent (Eds.), *Proceeding of the 20th Conference of the International Group for the Psychology of Mathematics Education* (Vol. 2, pp. 113–120). Valencia, Spain: PME.

Giacardi, L. (2003). I manuali per l'insegnamento della geometria elementare in Italia fra Otto e Novecento. In G. Chiosso (Ed.), *Teseo: Tipografi e editori scolastico-educativi dell'Ottocento* (pp. 97–123). Milan: Editrice Bibliografica.

Giacardi, L. (2012). Federigo Enriques (1871–1946) and the training of mathematics teachers in Italy. In S. Coen (Ed.), *Mathematicians in Bologna 1861–1960* (pp. 209–275). Basel: Birkhäuser.

Guala, E., & Boero, P. (2017). Cultural analysis of mathematical content in teacher education: The case of elementary arithmetic theorems. *Educational Studies in Mathematics, 96*(2), 207–227. https://doi.org/10.1007/s10649-017-9767-2.

Guo, S. (2010). *Chinese history of science and technology.* Beijing: Science Press (in Chinese).

Habermas, J. (1998). *On the pragmatics of communication.* Cambridge, MA: MIT Press.

Huang, R., Fan, Y., & Chen, X. (2017). Chinese lesson study: A deliberate practice, a research methodology, and an improvement science. *International Journal for Lesson and Learning Studies, 6*(4), 270–282.

Jaworski, B., Bartolini Bussi, M. G., Prediger, S., & Novinska, E. (2015). Cultural contexts for european research and design practices in mathematics education. In K. Krainer & N. Vondrová (Eds.), *CERME 9—Ninth congress of the European society for research in mathematics education* (pp. 7–35). Prague: ERME.

Leung, A., & Lopez-Real, F. (2002). Theorem justification and acquisition in dynamic geometry: A case of proof by contradiction. *International Journal of Computers for Mathematical Learning, 7*(2), 145–165.

Leung, F. K. S., Graf, K. D., & Lopez-Real, F. J. (Eds.). (2006). *Mathematics education in different cultural traditions. A comparative study of East Asia and the West. The 13th ICMI study.* New York: Springer.

Leung, F. K. S., Park, K., Shimizu, Y., Xu, B. (2015). Mathematics education in East Asia. In S. J. Cho (Ed.), *The Proceedings of the 12th International Congress on Mathematical Education: Intellectual and Attitudinal Challenges* (pp. 123–144). Basel: Springer.

Linati, P. (2011). *L'algoritmo delle occasioni perdute. La matematica nella scuola della seconda metà del Novecento* [A historical overview on proposals of reforms in the teaching of mathematics in Italy]. Trento: Erickson.

Malara, N. A., & Zan, R. (2008). The complex interplay between theory and practice: Reflections and examples. In L. English (Ed.), *Handbook of international research in mathematics education* (2nd ed., pp. 539–564). New York: Routledge.

Marchi, M. V., & Menghini, M. (2013). Italian debates about a modern curriculum in the first half of the 20th century. *The International Journal for the History of Mathematics Education, 8*, 23–47.

Mariotti, M. A. (2009). Artefacts and signs after a Vygotskian perspective: The role of the teacher. *ZDM Mathematics Education, 41*, 427–440.

Mariotti, M. A. (2012). ICT as opportunities for teaching-learning in a mathematics classroom: The semiotic potential of artefacts. In T. Y. Tso (Ed.), *Proceedings of the 36th Conference of the International Group for the Psychology of Mathematics Education* (Vol. 1, pp. 25–45). Taipei, Taiwan: PME.

Martinez, M., & Pedemonte, B. (2014). Relationship between inductive arithmetic argumentation and deductive algebraic proof. *Educational Studies in Mathematics, 86*(1), 125–149.

Mellone, M., & Ramploud, A. (2015). Additive structure: An educational experience of cultural transposition. In X. Sun, B. Kaur, & J. Novotná (Ed.), *Proceedings of the 23rd ICMI Study: Primary Mathematics Study on Whole Numbers* (pp. 567–574). Macao: University of Macao.

Mellone, M., Ramploud, A., Di Paola, B., & Martignone, F. (2017). Cultural transposition as a theoretical framework to foster teaching innovations. In B. Kaur, W. K. Ho, T. L. Toh, & B. H. Choy (Eds.), *Proceedings of the 41st Conference of the International Group for the Psychology of Mathematics Education* (Vol. 1, p. 244–51). Singapore: PME.

MIUR. (2010). *Progetto Lauree Scientifiche.* http://hubmiur.pubblica.istruzione.it/web/universita/progetto-lauree-scientifiche.

MIUR. (2012). *Indicazioni Nazionali per il Curricolo della scuola dell'infanzia e del primo ciclod'istruzione.* http://hubmiur.pubblica.istruzione.it/alfresco/d/d/workspace/SpacesStore/162992ea-6860-4ac3-a9c5-691625c00aaf/prot5559_12_all1_indicazioni_nazionali.pdf.

MIUR, SIS & UMI-CIIM. (2001). *Matematica 2001.* Resource document. http://www.umi-ciim.it/materiali-umi-ciim/primo-ciclo/.

Mok, I. A. C. (2013). Five strategies for coherence: Lessons from a Shanghai teacher. In Y. Li & R. Huang (Eds.), *How Chinese teach mathematics and improve teaching* (pp. 120–133). New York: Routledge.

Pedemonte, B. (2007). How can the relationship between argumentation and proof be analysed? *Educational Studies in Mathematics, 66*, 23–41.

Pedemonte, B., & Balacheff, N. (2016). Establishing links between conceptions, argumentation and proof through the ck¢-enriched Toulmin model. *Journal of Mathematical Behavior, 41*, 104–122.

Pedemonte, B., & Buchbinder, O. (2011). Examining the role of examples in proving processes through a cognitive lens. *ZDM Mathematics Education, 43*(2), 257–267.

Pedemonte, B., & Reid, D. (2010). The role of abduction in proving processes. *Educational Studies in Mathematics, 76*(3), 281–303.

Severi, F. (1932). Didattica della matematica. *Enciclopedia delle enciclopedie: Pedagogia, Formiggini* (pp. 362–370). Rome: A. F. Formiggini.

Sun, X. (2015). Chinese core tradition to whole number arithmetic. In X. Sun, B. Kaur, & J. Novotná (Eds.), *Proceedings of the 23rd ICMI Study: Primary Mathematics Study on Whole Numbers* (pp. 140–148). Macao, China: University of Macao. http://www.umac.mo/fed/ICMI23/proceedings.html. Accessed 20 Feb 2016.

Sun, X. (2016a). *Uncovering Chinese pedagogy: Spiral variation—The unspoken principle for algebra thinking to develop curriculum and instruction of "TWO BASICS".* Paper presented at 13th International Congress on Mathematical Education (ICME-13), Hamburg. http://www.icme13.org/files/abstracts/ICME-13-Invited-lectures-Sun.pdf.

Sun, X. (2016b). Spiral variation: A hidden theory to interpret the logic to design Chinese mathematics curriculum and instruction in mainland China [in Chinese]. *World Scientific Publishing C.P.* https://doi.org/10.1142/9789814749893.

Sun, X., Kaur, B., & Novotna, J. (Eds). (2015). *Conference proceedings of the twenty-third ICMI study: Primary mathematics study on whole numbers.* Macao: University of Macao.

Toulmin, S. E. (1958). *The uses of argument.* Cambridge: Cambridge University Press.

Vianna, E., & Stetsenko, A. (2006). Embracing history through transforming it: Contrasting Piagetian versus Vygotskian (activity) theories of learning and development to expand constructivism within a dialectical view of history. *Theory & Psychology, 16*(1), 81–108.

Waschescio, U. (1998). The missing link: Social and cultural aspects in social constructivist theory. In F. Seeger, J. Voigt, & U. Waschescio (Eds.), *The culture of the mathematics classroom* (pp. 221–241). Cambridge: Cambridge University Press.

Wertsch, J. V. (1991). *Voices of the mind: A sociocultural approach to mediated action.* Cambridge, MA: Harvard University Press.

Zhang, J. (2016). *The reform and development of plane geometry middle school course in China.* Paper presented at 13th International Congress on Mathematical Education (ICME–13), Hamburg. http://www.icme13.org/files/abstracts/ICME-13-invited-lectures-Zhang.pdf.

Chapter 5
The German Speaking Didactic Tradition

Rudolf Sträßer

Abstract This chapter gives a historical sketch on the development of the Didactics of Mathematics as practised in German language countries, from the nineteen sixties to the present. Beginning from "Stoffdidaktik", anecdotal teaching episodes and large scale psycho-pedagogical research, Didactics of Mathematics has developed into a well defined scientific discipline. This tradition has embraced a plethora of areas researched via a range of methodological approaches stretching from local case studies to large scale surveys in order to capture broad perspectives on research into teaching and learning mathematics. Enlarged "Stoffdidaktik", and Didactics of Mathematics as a "design science", draws on a large number of classroom studies looking into a wide variety of specific aspects of teaching and learning mathematics. The sometimes politically driven, large-scale studies use sophisticated statistical methods to generate generalizable results on the state of educational systems, especially in terms of the changes and trends over time.

5.1 Introductory Remark

It is impossible to give adequate credit to all important contributions to Didactics of Mathematics (internationally often called mathematics education or research in mathematics education) in the German speaking countries in a short text of twenty pages. This chapter attempts to provide an overview of major contributions to the development of the Didactics of Mathematics in German language countries in past seven decades. As such, this chapter is unavoidably personally coloured and the author sincerely apologises to those who have not been mentioned or been misrepresented.

Supported by Stefan Krauss and Kerstin Tiedemann and comments from Barbro Grevholm, Edyta Nowinska and Nad'a Vondrowa

R. Sträßer (✉)
University of Giessen, Giessen, Germany
e-mail: Rudolf.straesser@math.uni-giessen.de

© The Author(s) 2019
W. Blum et al. (eds.), *European Traditions
in Didactics of Mathematics*, ICME-13 Monographs,
https://doi.org/10.1007/978-3-030-05514-1_5

This chapter first gives a historical sketch on German-speaking Didactics of Mathematics starting in the nineteen sixties. We then give an overview on the situation of Didactics of Mathematics (or research on mathematics education) in the 21st century until present in these countries. The paper finishes with a personal perspective (Rudolf Sträßer alone) on the future of German speaking Didactics of Mathematics. As an additional perspective from outside the German-speaking countries, three researchers from the Nordic countries (Norway/Sweden), from Poland and from the Czech Republic comment on interactions between research in their own countries and the German-speaking Didactics of Mathematics. A more detailed picture on German-speaking Didactics of Mathematics could be found in a parallel meeting during the Thematic Afternoon of ICME-13 (Hamburg Germany), where a whole afternoon was totally devoted to Mathematical Didactics from the perspective of the host country of the conference. The table of contents of a book, which emerged out of this activity (see Jahnke & Hefendehl-Hebeker, 2019) is added as an appendix at the end of this paper, representing the plans for the book as of June 2018.

5.2 Historical Sketch on German Speaking Didactics of Mathematics

5.2.1 Starting Point in the 1960s

Even if a historical sketch starts with the nineteen sixties, it is more than appropriate to give credit to work in Didactics of Mathematics already existing in these years in the German speaking countries (for another detailed description of Didactics of Mathematics in German speaking countries see e.g. Schubring, 2014, for a text in German nearer to the past see Steiner 1978).

In the nineteen sixties, three types of texts were available on the teaching and learning of mathematics in German speaking countries: First, personal reports from mathematics classrooms, which were given by experienced teachers and personnel from the education administration, often combined with document analysis for curriculum development purposes. A typical exemplar of this genre are the reports for the renewed International Commission on Mathematics Instruction (ICMI) like Drenckhahn (1958) or Behnke and Steiner (1967).

Second, German speaking Didactics of Mathematics at that time was strong in the so-called Subject Matter Didactics ("Stoffdidaktik"). "Stoffdidaktik" can be described as the mathematical analysis of the subject matter to be taught. The purpose of the analysis was to find appropriate ways or even the best (in certain approaches: one and only) way to make a mathematics topic accessible and understandable for students. This work was completed by mathematicians, teacher trainers, textbook authors and mathematics teachers—and often authors of "Stoffdidaktik" had more than one of these roles to play. Two strands of "Stoffdidaktik" can be distinguished, namely the one done by university and "Gymnasium" teachers as separated from the one done by teacher trainers for primary and general education. Even if separated from an institutional point of view, the common basic methodology is mathematics as a scientific discipline together with anecdotal classroom experience. Differences of the two strands come down to different mathematical topics for teaching according to the two—at that time in Germany—clearly different types of schools, namely Gymnasium (at that time targeting a small minority of students heading for university) and "Volksschule" (the then standard type of schooling for eight years aiming at the vast majority of students in general education ending at a student age of about 15 years). Typical exemplars of this genre are Oehl (1962) for "Volksschule" or the famous work by Lietzmann aimed at teaching in Gymnasium. The first edition of Lietzmann was published in 1916, had intermediate editions during the Weimar republic and the Nazi regime, and was re-published with the same title in 1951. The latest amended edition was published in 1985 under the editorship of Jahner (see Jahner 1978/1985).

The third genre is quantitative, mainly comparative studies, which were often completed by scientists from university departments of psychology. This research analysed the psychological preconditions and constraints of the learners. In Germany, these works had a long tradition from Katz (1913) (already sponsored by ICMI) to Strunz (1962) (4th edition), but did not develop into a strong research paradigm in the German speaking countries (for this judgement see Schubring, 2012).

5.2.2 Institutionalisation

In the second half of the nineteen sixties, in connection with the 'sputnik crisis', one German societal debate circled around a so-called educational catastrophe ("Bildungskatastrophe") within the West-German Federal Republic. As a consequence and to foster economical growth, a social and political move made efforts to expand the educational system in the Federal Republic of Germany, prolonging the time of compulsory schooling and especially promoting the scientific school subjects including mathematics. On a societal level, this led to the creation of new universities, the training of more mathematics teachers, more teacher training for mathematics and the academisation of teacher training for primary education teachers. A detailed description of these developments can be found in Schubring (2016, pp. 15ff).

For Didactics of Mathematics as a scientific discipline, this societal change had important and long lasting consequences: For the first time in German history, full professorships in Didactics of Mathematics were created at universities—especially within newly founded institutions. In 1975, professionals in Didactics of Mathematics founded a scientific society for Didactics of Mathematics, the "Gesellschaft für Didaktik der Mathematik (GDM)", followed by the creation of a new research journal in 1980, the "Journal für Mathematik-Didaktik (JMD)"—still under publication and now (since 2010) with Springer. On the initiative of the Volkswagen Foundation in 1972, a research institute for Didactics of Mathematics was founded at Bielefeld University, the "Institut für Didaktik der Mathematik (IDM)". This was eventually integrated (1999) into the Faculty of Mathematics at Bielefeld University and gradually turned into a "normal" university institute with a "standard" mission in research and teacher training, but still one of the best libraries specialising in Didactics of Mathematics worldwide. A different indicator of the institutionalisation of German speaking Didactics of Mathematics was the organisation of the third International Conference on Mathematics Education (ICME-3), which was held in Karlsruhe in 1976 with a strong participation from German speaking scholars in the preparation and the activities of this conference (see Athen, 1977 for the original proceedings of the meeting).

5.2.3 The 1970s/1980s: The "Realistic Turn"

Together with the institutionalisation of Didactics of Mathematics as a scientific discipline, research on the teaching and learning of mathematics in schools underwent a major change. Compared to the situation in the nineteen sixties, we can identify two non-convergent developments, which drastically changed the research landscape in West-Germany and Austria: On the one hand, we see a "realistic turn" with more detailed empirical research, less document analysis and anecdotal reports, which is done to develop a description and (if possible: a causal) explanation of what is going on in the teaching and learning of mathematics. This development of Didactics of Mathematics followed a scientific development in Germany, which—at that time—started to favour empirical research over philosophical and historical research in pedagogy. Internally, inside Didactics of Mathematics, this move was a scientific answer to the failure of the 'Modern Math Movement'. Taking set theory as the foundation of Mathematics in schools together with an axiomatic approach to teaching, abstract algebra and a high importance of logic in schools became less prominent at that time. In general, research in the 1970/80 s favoured small scale, qualitative empirical research, looking at transcripts of classroom communication and interviews viewed through a specific theoretical lens. This was done with a focus on the communication process, not so much on the mathematical achievements of the students in order to identify the opportunities and challenges associated with interactions in mathematics classrooms. The move to a more empirical, small scale approach in research is accompanied by a methodological development, which favours qualitative or sometimes linguistic analysis of classroom processes, as opposed to a more traditional experimental research approach. In West-Germany, this movement was initiated by the Bauersfeld group from the Bielefeld Institute and implied a "turn to the everyday classroom" (for a more detailed description of this change of paradigms see Voigt, 1996, pp. 383–388).

   To sum up the situation of Didactics of Mathematics in Austria, West-Germany and German-speaking Switzerland, this discipline started as a rather homogeneous field made up of subject matter didactics and classroom studies. In the 1980s, it diversified into a plethora of research with different paradigms and on a variety of aspects of the teaching and learning (process) of mathematics. In 1992, Burscheid, Struve

& Walther analysed the topics of the "publications in West German professional journals and research reports of the IDM", the research institute at Bielefeld university mentioned above and came to the following important research areas (see Burscheid, Struve, &Walther, 1992, p. 296):

- "empirical research
- subject matter didactics
- applications in mathematics teaching
- historical and philosophical investigations
- methodological aspects of mathematics education
- principles of mathematics education
- the epistemological dimension of mathematics education
- proving."

Besides the methodological decision to give empirical research a greater importance, research not only in subject matter didactics concentrated on Arithmetic and Calculus—with a relative neglect of topics from Geometry. With 'Modern Math' (logically based on set theory) becoming less important, the role of applications in teaching Mathematics came into the centre of one German speaking strand of Didactics of Mathematics with a focus on the modelling circle (see Pollak, 1979). Historical and philosophical investigations served as a basis for enriching classroom teaching of Mathematics with information on its role in culture and history. Principles of mathematics education (like the genetic principle to follow the historic line of development of mathematical topics or the operative principle to learn through close inspection of operations on carefully selected situations) were identified as a way to structure the variety of school mathematics topics to form a coherent whole, if possible showing the general value of Mathematics for education as a whole.

In the 1970/80s, discussion was focused on the idea of a comprehensive theory of Didactics of Mathematics taking into account its epistemological dimensions. In 1974, Hans-Georg Steiner, one of the directors of the institute at Bielefeld University, edited a special issue of Zentralblatt für Didaktik der Mathematik (ZDM) with contributions from Bigalke, Freudenthal, Griesel, Otte and Wittmann (among others), looking into the possibility of a comprehensive theory of Didactics of Mathematics, especially ideas on its subject area, its scientific character and its relation to reference disciplines, especially mathematics, psychology and educational science. No coherent conceptualisation was reached. About ten years later in 1983, a comparable debate in the Journal für Mathematikdidaktik (JMD) with protagonists Burscheid, Bigalke, Fischer and Steiner also did not produce a coherent paradigm for Didactics of Mathematics, but ended with the conclusion that Didactics of Mathematics may never develop into a science based on a single paradigm underpinned by one and only one unifying theoretical approach (for details see the chapter on "Theories of and in mathematics education related to German speaking Countries" in the volume edited by Jahnke et al., 2019). In addition to these discussions, in 1983, Steiner took the initiative of creating an international group on "Theory of Mathematics Education (TME)", which had a first meeting after ICME-5 in Adelaide in 1984. This international group had four follow-up conferences in different locations, but also could not arrive at a joint unifying conceptualisation of Didactics of Mathematics as a scientific discipline. The international work on a theory of Didactics of Mathematics was complemented by a series of bi-lateral symposia organised by Steiner (see the French-German symposium in 1986, the Italian-German seminars in 1988 and 1992 and the International Symposia on Mathematics Education in Bratislava in 1988 and 1990; for all activities see Biehler & Peter-Koop, 2007, p. 27f). By the late 1990s, this line of discussion began to wane, at least in the German speaking countries, but re-emerged internationally in the 21st century with the meta-theoretical approach of networking theories (for a comprehensive presentation of this movements see Bikner-Ahsbahs & Prediger, 2014).

5.2.4   Didactics in the German Democratic Republic (GDR)

The attentive reader will have noticed that in some places, we employed the wording "West-Germany"—and this was done on purpose. Some reports on the history of Didactics of Mathematics in Germany tend to forget that—because of political reasons and differences—research on the teaching and learning of mathematics developed differently in West-Germany, the Federal Republic of Germany ("Bundesrepublik Deutschland—BRD") and in East-Germany, the German Democratic Republic ("Deutsche Demokratische Republik—DDR"). This was even marked by a different designation of the activity. In East Germany, the science of teaching and learning mathematics was called "Methodik" (direct, but misleading translation: "methodology"). For more detailed information see Henning and Bender (2003). If we follow the basic and informative text by Bruder (2003), four characteristics have to be mentioned for "Methodik" in the German Democratic Republic (GDR):

(1) As usual in a "socialist" country, the science of teaching and learning mathematics was highly controlled by political authorities, which tried to implement a uniform planning for the comprehensive school installed in the country. This implied the use of one and only one textbook in the GDR, accompanied by just one set of teaching aids.

(2) Teaching and learning of mathematics had a systematic disciplinary orientation following the systematisation of the discipline Mathematics, but trying to cope with the teaching reality. Research into teaching and learning picked up the difficulties experienced by teachers in classrooms and focussed on intervention to overcome these difficulties. A theory of teaching/learning played a minimal role in the creation of textbooks, teaching aids and detailed suggestions for the way to teach mathematics.

(3) Periodical repetition and mental training of basics were the backbone of the optimization of instruction and construction of learning environments. The overall learning goal was deeply rooted in the ideas of a developed social personality, i.e. a socialist idea of humankind and a high esteem of mathematics and science in the socialist society. This "Methodik" was highly accepted by teachers

due to the institutionalisation of these ideas by a high proportion of lessons on Didactics of Mathematics at university and in practical training at school during the study and no inconsistencies between pre-service teacher education and in-service teaching experience.

(4) The teaching in classroom was marked by a linear, uniform structure of subjects to be taught. Subjects (like mathematics) had the priority compared to the individual needs of the students—with the consequence that students' differences were taken into account by means of inner differentiation in a uniform educational system. For gifted students, special support was available from Mathematical Olympiads on different levels and from special schools ("Spezialschulen") for students who seemed to be able to cope with higher demands and extended training in areas like sports, music or mathematics.

These ideas and practices of a "socialist" (mathematics) education were left behind with the fall of the German Democratic Republic in 1989. Nevertheless, careful and attentive observers of the educational systems in Germany can still recognise consequences of these traditions in some regions of the former German Democratic Republic such as a higher importance given to mathematics and science in the former German Democratic Republic.

5.2.5   1990s: The PISA Shock

During the 1990s, German Didactics of Mathematics had to cope with the change induced by having only one state in the German country and the gradual adaptation of the East German regions to the educational system of West-Germany. In this period, an external incident drastically changed the public views on teaching and learning mathematics in German-speaking countries. During 1996, the results of the Third International Mathematics and Science Study (TIMSS) were released and showed Germany to be a nation with an average student achievement in grades 7 and 8 if compared to all nations participating in this study. Two years later, the Programme for International Student Assessment (PISA) placed Germany even below average for the results in mathematical literacy inducing what was called in Germany the *PISA-shock*. The realisation that German (mathematics) education in schools was not performing to expectations had numerous political consequences. Most prominent were policy sponsored efforts to enhance the teaching and learning especially of mathematics in general education by means of programs to enhance this educational effort. The "SINUS" study (described further down in Sect. 5.3.3) was the most important effort of this type of government sponsored intervention aiming at professional development of mathematics teachers. Another consequence of the *PISA-shock* were studies devoted to the identification of teaching standards in Germany (heavily drawing on a concept of teaching/learning mathematics with the intention to offer mathematical *Bildung*, e.g., Winter, 1995) and numbers of (regional and national) evaluations aimed at gathering detailed information on the achievement (or not) of students. Politicians also hoped to get pertinent information and ideas how to enhance the knowledge and skills of the population to do better economically (for details see below in Sect. 5.3.3).

Additional activities originating from the *PISA-shock* were large scale projects from the regional education ministries aimed at increasing teaching quality in "normal" classrooms. The regional political entities responsible for education, the „Länder" joined forces to set up the qualitative large scale development study „SINUS" and later its follow-up study „SINUS-transfer" (for the background and detailed information see http://www.sinus-transfer.eu). The ministers of education in the „Länder" of the Federal Republic of Germany agreed upon 11 „modules", that is, 11 topics for enhancing the teaching quality in mathematics and science: developing a task culture, scientific working, learning from mistakes, gaining basic knowledge, cumulative learning, interdisciplinary working, motivating girls and boys, cooperative learning, autonomous learning, progress of competencies, quality assurance (see http://www.sinus-transfer.eu/). The two development projects SINUS and SINUS-transfer aimed at helping the classroom teacher enhance her/his teaching by developing examples of good teaching and were heavily supported by research institutes and didacticians. The most important activity of the projects consisted of the organisation of teacher cooperation and the dissemination of good teaching units developed to illustrate the goals of the 11 modules.

5.3   The 21st Century—At Present

By the 21st century, German speaking Didactics of Mathematics was well-established as a research field. An indication of its strength was the invitation to host a second International Congress on Mathematics Education, following ICME-3, namely ICME-13 in Hamburg in 2016. Being able to structure and manage a congress of about 3500 participants from 107 countries demonstrates the maturity and stability of the German-speaking community of didacticians of Mathematics. In addition, scholars from the German speaking community have a history of support, through their time and expertise, for European and international research efforts in Mathematics Education. Numerous colleagues from German speaking countries serve in the steering bodies of European and international research organisations of Didactics of Mathematics and in the organisation of working groups in research organisations such as the International Conference on Mathematics Education (ICME), the group on Psychology of Mathematics Education (PME), the European Society for Research in Mathematics Education (ERME) or the International Conferences on the Teaching of Mathematical Modelling and Applications (ICTMA)—to name just a few.

The research situation in German speaking Didactics of Mathematics can be characterised by three major strands of research, which for the second case offer a broad variation even within this strand. We distinguish a strand of enlarged Stoffdidaktik, which now also embraces the design of learning environments (sometimes controlled by learning theories; for details see 5.3.1 below). This is different from a plethora of (mostly qualitative) descriptive and for the most part comparative case studies, especially "classroom studies". A third strand is a variety of large scale developmental or evaluation studies, which—following a quantitative or qualitative approach—are sometime deeply influenced by the TIMSS or PISA activities and—for the majority—are also motivated, if not sponsored by (educational) politicians. For a more detailed account on Didactics of Mathematics in German-speaking countries see Jahnke et al. (2019), a book describing the parallel Thematic Afternoon on German-speaking Didactics of Mathematics, which was held in parallel to the afternoon on European traditions. An appendix in the end of this text offers the table of contents of this book.

5.3.1   Stoffdidaktik Enlarged—The Design of Learning Environments

The first major strand of German-speaking Didactics of Mathematics developed out of traditional Stoffdidaktik (based on the mathematical analysis of the subject matter to be taught). In addition to the diligent analysis of mathematical contents, enlarged Stoffdidaktik takes into account additional influences on the subject matter taught, especially the history and epistemology of mathematics, fundamental ideas of mathematics (if visible), and information on the learner and her/his pre-requisites for learning (including 'basic mental models' and beliefs of the learner). This approach also embraces empirical studies on consequences of subject matter innovation and can be distinguished from traditional Stoffdidaktik by its acceptance of a developed cooperation with other disciplines (e.g. Educational Psychology) looking into the teaching and learning of mathematics. On the occasion of ICME-13, the German-speaking research journal Journal für Mathematikdidaktik (JMD) published a supplement totally devoted to modern Stoffdidaktik (see https://link.springer.com/journal/13138/37/1/suppl/page/1) with an introductory description of this development.

Growing out of the search of Stoffdidaktik for the best way to teach mathematics in an understandable way, some German-speaking didacticians make the design of learning environments the defining 'kernel' of research in mathematics education (Didactics of Mathematics). One of the protagonists of this approach, Erich C. Wittmann, came up with the definition of "Didactics of Mathematics as a design science" (see already the title of Wittmann, 1992/1995). Wittmann starts from the distinction of a "core" of mathematics education and "areas related to mathematics education" (like mathematics as a scientific discipline, sociology, pedagogy and general didactics; see Fig. 5.1 on p. 357 of Wittmann 1995). For mathematics education, he identifies "the framework of a design science" as "a promising perspective for fulfilling its tasks and also for developing an unbroken self-concept of mathematics educators" (loc. cit., p. 362). As a design science, "the core of mathematics education concentrates on constructing *artificial objects*, namely teaching units, sets of coherent teaching units and curricula as well as the investigation of their possible effects in different educational "ecologies"" (loc. cit., p. 362f).

5.3.2   Ongoing Diversification of Classroom Studies

The second major strand of German-speaking Didactics of Mathematics can be globally described as Classroom Studies. This strand is very diverse and often done in cooperation with other disciplines, such as psychology, pedagogy or educational sciences. These studies can be described as mostly qualitative and descriptive case studies, which seek to reconstruct specific aspects of everyday teaching and learning. For the moment, there is no comprehensive overview of this research field, but an impression nonetheless can be gleaned from the following list of typical topics (compiled from the last five years of the Journal für Mathematikdidaktik JMD):

- the use of technology in mathematics teaching and learning
- subject matter analysis—proof and argumentation
- modelling in mathematics classrooms
- the role of language
- early childhood and primary education
- variables of good class management
- gender and teaching & learning mathematics
- textbook research
- history and epistemology of mathematics
- semiotics and mathematics
- teacher training (preparatory and in-service)
- competencies in mathematics.

One way to develop an understanding of *Classroom Studies* is to learn more about their aims, history and ways of working. First of all, *Classroom Studies* can be characterized by their *sociological orientation* on learning mathematics. Often, the focus is on the social dimension of learning processes; not so much to understand one individual case, but to compare a number of cases in order to identify common features and differences on a theoretical level. It is because of this comparative approach that researchers from that strand ascribe a meaning to their results that goes well beyond the individual case.

The strand of *Classroom Studies* was initiated by the works of *Heinrich Bauersfeld* and his colleagues in the 1980s. Their key assumption was that mathematical knowledge is always developed within social interaction (social negotiation of mathematical meaning). First, they observe everyday mathematics classes and collect different kinds of data, such as videos, audiotapes or written products. Second, they transcribe the processes of social interactions. Third, they analyse these transcripts by means of the interaction analysis. This method combines a sociological and a mathematical perspective. It aims at reconstructing the thematic development of a given face-to-face interaction. A very important step in this analytical work, then,

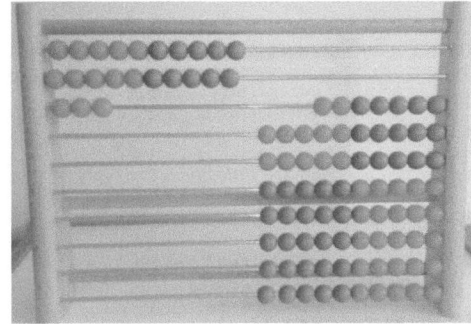

Utterance 1 on 32:

"because here are 3 and here are 2"

Utterance 2 on 15:

"one row of tens and five beads"

Fig. 5.1   The bead frame

is to compare the interpretations of different scenarios and to identify similarities and differences. As a result, the researchers finally describe different types of a phenomenon. And these descriptive results can be used to analyse other lessons or to develop suggestions for improvement.

To illustrate the strand of Classroom Studies, we give a short *example* from a research project about primary students and their development of subject-related language that was presented during the Thematic Afternoon of ICME-13 by Kerstin Tiedemann.

Research Example 1 (by Kerstin Tiedemann):

Material artefacts and classroom communication

The example is about a young girl named Hanna. She is 9 years old, goes to school in the third grade, and speaks German as her first language. For that reason, her utterances (utterance 1-5) are translated from German to English. Hanna was filmed in discussions with her teacher Britta over a period of three months. In their lessons, Hanna and Britta work with two different didactic materials one after the other: with a bead frame and with Dienes blocks.

First, they work with the bead frame (see Fig. 5.1).

Hanna uses quite unspecific language (Fig. 5.1, utterance 1). She refers only to amounts, but she does not mention what the objects are that she is talking about. For example, she names a twenty-three (23) on the bead frame as a thirty-two (32) and gives the argument: "because this is 3 and this is 2". At this point, tens and ones are exchangeable for Hanna. But, then, in interaction with Britta, she starts to specify her language in relation to the bead frame (utterance 2). She learns to distinguish tens and ones and refers to them as 'rows of tens' and single 'beads'. For example: "1 row of tens and 5 beads."

Later on, Hanna and Britta work with the Dienes blocks (see Fig. 5.2), with another material.

In those lessons—and this is really interesting -, Tiedemann (2017) could reconstruct a surprisingly analogue language development. First, Hanna uses quite unspecific language, refers only to amounts and often mixes up the tens and ones (Fig. 5.2, utterance 3). But then, she starts to specify her language again, this time in relation to the Dienes blocks (utterance 4). She learns to distinguish tens and ones and refers to them as 'bars of tens' and single 'cubes'. It is this specification of language that allows Hanna to compare both materials in the end and to describe their mathematical similarity (utterance 5). With regard to a forty-three

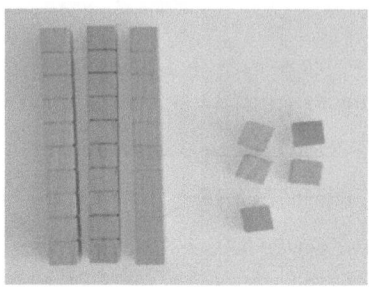

Utterance 3 on 53:

"because here are five and here are three"

Utterance 4 on 25:

"two bars of tens and five little cubes"

Utterance 5 on 42:

"for the four tens, you take rows and not bars, okay?"

Fig. 5.2 Dienes' blocks

(43) on the bead frame, she says in comparison to the Dienes blocks: "For the 4 tens, you take rows and not bars, okay?"

To sum up, Hanna is a prototypical example of a certain type of language development – she begins with an unspecific language use, but then she specifies her language while interacting with Britta and is then able to compare different representations of tens and ones. It is in this process that she shows more and more understanding of the mathematical concept of place value.

5.3.3   Large Scale Comparative Studies

The third major strand in German-speaking Didactics of Mathematics grew in large parts out of the *PISA-shock* mentioned in Sect. 5.2.5. Politicians and researchers turned the deception on the (average or below average) German results in the TIMSS and PISA-studies into efforts to know better and with the help of statistical inquiry about the actual achievements of students when learning mathematics in general education. This created the opportunities and needs for quantitative large scale evaluation studies not only in mathematics, but also in German, English and Science teaching and learning. First, the deceiving results from the TIMSS- and the TIMSS-video-study led to German participation in the PISA-studies, where German didacticians of mathematics took a major role in preparing the research instruments. This was complemented by the creation of an extension of the original PISA-study (called PISA-E), which allowed for an intra-national comparisons of the teaching/learning practice of individual German *Länder* (the political regional entities below the whole nation, which—by constitution—are responsible for education in general). From the PISA-study in 2003 onwards, we can see a heavy involvement of German didacticians of mathematics in such comparative evaluation studies on the teaching and learning of mathematics in German general education (for details see http://archiv. ipn.uni-kiel.de/PISA/pisa2003/fr_reload_eng.html?mathematik_eng.html). To illustrate this development, we present an evaluative study on teacher competence, which was embedded in the PISA process, the COACTIV-study on "teacher competence as a key determinant of instructional quality in mathematics" and COACTIV-R (a continuation of COACTIV from 2007 onwards) on "teacher candidates acquisition of professional competence during teaching practice" (for details see → https://www. mpib-berlin.mpg.de/coactiv/index.htm). This example was also presented during the Thematic Afternoon at ICME-13.

Research example 2 (by Stefan Krauss): The Impact of Professional Knowledge on Student Achievement

The German COACTIV 2003/04 research program (*Co*gnitive *Activ*ation in the Classroom: Professional Competence of Teachers, Cognitively Activating Instruction, and Development of Students Mathematical Literacy) empirically examined a large representative sample of German secondary mathematics teachers whose classes participated in the German PISA study and its longitudinal extension during 2003/04. Since the students of the Grade 9 classes which were tested in PISA 2003 in Germany were examined again in Grade 10 in the following year, the design allowed to analyse the impact of teacher competencies (as assessed by COACTIV) on students' learning gains in mathematics (as assessed by PISA-03 and - 04). The COACTIV-project was funded by the German research foundation (DFG) from 2002 to 2006 (directors: Jürgen Baumert, Max-Planck-Institute for Human Development Berlin; Werner Blum, University of Kassel; Michael Neubrand, University of Oldenburg). The rationale of the project and its central results are summarized in the compendium by Kunter, Baumert, Blum, et al. (2013a).

Analyses of structural equation models revealed that the newly developed test on teachers' pedagogical content knowledge (PCK) was highly predictively valid for the lesson quality aspects of cognitive activation and individual learning support. Furthermore, a "black-box"-model for the direct impact of teacher competence on students learning gains (without mediation by aspects of instructional quality, adapted from Kunter et al. 2013b) shows a regression coefficient of 0.62 of PCK on the students' learning gain and can for instance be interpreted in the sense that classes of teachers who scored one standard deviation *below* the PCK-mean of all teachers yielded an average learning gain within one school year (in terms of effect sizes) of about $d = 0.2$ and classes of teachers who scored one standard deviation *above* the PCK-average yielded a learning gain of about $d = 0.5$. Since the average learning gain of all PISA-classes was $d = 0.3$, this demonstrates that a difference of two standard deviations in the PCK of a mathematics teacher can make a difference in the learning gain of the typical amount of what on average is learned in a whole school year. The construction of a psychometric test on the PCK of mathematics teachers meanwhile inspired studies on other school disciplines like German, English, Physics, Latin, Musics and for religious education (see Krauss et al., 2017).

Besides evaluation studies linked to PISA, other major large scale comparative studies were hosted by the International Association for the Evaluation of Educational Achievement (IEA), which had already managed the TIMSS- and TIMSS-video-study. A major activity in German-speaking countries was the Teacher Education and Development Study on Mathematics named TEDS-M (for details see Blömeke et al., 2014, for more information on the study see https://arc.uchicago.edu/reese/projects/ teacher-education-and-development-study-mathematics-teds-m). Inspired by ideas from Shulman (1986/1987), the TEDS-M-study looked into the "Policy, Practice, and Readiness to Teach Primary and Secondary Mathematics in 17 Countries" (quote from the subtitle of the technical report Tatto 2013) including Germany. As described by Tatto (2013) "The key research questions focused on the relationships between teacher education policies, institutional practices, and future teachers' mathematics and pedagogy knowledge at the end of their pre-service education." (from the back-cover of Tatto, 2013; for results see Tatto et al., 2012; a detailed presentation of results in German can be found in Blömeke, Kaiser, and Lehmann 2010a, b). In Germany, the international TEDS-M study resulted in several further national follow-up studies, amongst others the study TEDS-LT, that compared the development of the professional knowledge of student teachers in the subjects mathematics, German language and English as first foreign language ("EFL", see Blömeke et al., 2013), or the TEDS-FU study that analysed situational facets of teacher competence (Kaiser et al., 2017). While TEDS-M and TEDS-LT used paper and pencil based assessment of the mathematical content (MCK), mathematical pedagogical content knowledge (MPCK), and general pedagogical knowledge (GPK) of teacher students and teachers, the instruments for measuring teacher competence have been expanded to include also video and online testing in TEDS-FU.

Another consequence of the search for information on students' achievement were a number of assessment studies (so-called Schulleistungsstudien) installed by the majority of the regions (*Länder*) responsible for general education. In nearly all regions detailed studies of the results of teaching and learning mathematics were set up on various school levels. According to regional decisions, they use different methods (like traditional textbook tasks, multiple choice questionnaires and other instruments) and they are not restricted to the school subject mathematics. Examples of these studies are the VERA study with a representative comparison of classes and schools in grade 3, for most regions also in grade 8, looking at least into the two school subjects Mathematics and German (for more information see https://www.iqb.hu-berlin.de/vera). A more local, but more comprehensive study using the VERA framework is the KERMIT-study in Hamburg. From 2012 onwards it looks into the teaching and learning of mathematics (together with other school subjects) in Grades 2, 3, 5, 6, 7, 8 and 9 with the aim of informing teachers about the strengths and weaknesses of their classes, in order to use this information to tailor the teaching to students' needs (for details see http://www.hamburg.de/bsb/kermit/).

5.4   About the Future of German-Speaking Didactics of Mathematics

After preparing the historical sketch on German-speaking Didactics of Mathematics and especially informed by interviews I have made with Lisa Hefendehl-Hebeker and Hans-Georg Weigand, I offer a personal look into what may be the developments and challenges for German-speaking Didactics of Mathematics in the future. Obviously and as a sort of unavoidable kernel, a comprehensive, epistemologically and historically well informed understanding of Mathematics is crucial, including a concept of the vocational and social rule of this scientific discipline. With Didactics of Mathematics focussing on teaching and learning, another crucial point and great challenge is how the human brain works. These two major issues would perhaps give a chance of developing a comprehensive theory of Didactics of Mathematics.

Such a theory also has to model the many different impacts on teaching and learning mathematics, like the impact of the personal development of the student, the impact of the social situation the learner lives in and how the individual brain works apart from invariant development variables. This theory also has to account for the setting of the student within the family s/he comes from, the situation in the classroom, the political situation of the school, and the situation in society at large. Because of the many students with different mother tongues and because we have a new consciousness of the relation between thinking and speaking, the role of language in teaching and learning mathematics will be another major issue in future research into the Didactics of Mathematics, together with the use of technology, which deeply influences mathematics teaching and learning. These areas of research will only be helpful for the everyday classroom if we find out more about characteristics of good classroom management.

In terms of research methodology, the last decades clearly show that empirical research is important, but we should take care to keep the balance between empirical, maybe statistical quantitative research, qualitative investigations and conceptual, theoretical work in Didactics of Mathematics. As for research based on quantitative methods, a mixed methods approach may be appropriate. This balance allows for different approaches from a simplified empirical paradigm. For example subject oriented analysis and design experiments and philosophical and epistemological discussions about questions appear to be effective guidelines for mathematics education today.

Using a more global perspective, one can find different initiatives to systematize or to synthesize different theoretical approaches, from all over the world. In Germany, researchers like Angelika Bikner-Ahsbahs and Susanne Prediger could be named in this respect. It is visible that even today, the overall result is not a unifying theory, but limited local relationships between different approaches. It is clear that we do not, as yet, have an overarching, comprehensive theory.

A different challenge for the future is and will be bringing the results of empirical investigations, including large scale investigations to the school. How to "scale up" (sometimes local, limited) findings, insights and suggestions to bring them to the learner, to the school, to the administration, to politicians—apart from advancing Didactics of Mathematics as a scientific discipline.

5.5   Comments from Critical Friends

5.5.1   Doing Empirical Research Differently: The Nordic and German Cases. A View from the Nordic Countries

(By Barbro Grevholm)

Reasons for Nordic connections to Germany

Germany is a neighbour to the Nordic area and can easily be reached via land from Denmark, or sea from Sweden, Norway, Finland, and Iceland. German language is often the second foreign language, after English, learnt in the Nordic area and it used to be the first foreign language. The linguistic connections are many and in historical times German for long periods was the spoken language in Stockholm, for example. The classical *Bildningsresorna* (in German: Bildungsreisen; an English explanation is very difficult because of the concept of *Bildung*, which does not exist in English) went to Germany for many young persons in the Nordic countries. One famous example is the young Norwegian mathematician Abel.

Some early examples of German influence

In 1920, Salomon Eberhard Henschen, a medical professor in Stockholm, published a book entitled *Klinische und Anatomische Beiträge zur Pathologie des Gehirns. 5. Teil. Uber Aphasie, Amusie und Akalkulie* [Papers on Brain Pathology from Clinical Research and Anatomy, part 5 on aphasia, amusia and acalculia; title translated by RS]. This book can be considered an early contribution to studies on dyscalculia. Another example is the Kassel-Exeter study, which had links to Norway and the KIM-study (for details see Streitlien, Wiik, & Brekke, 2001) and there are also later master studies, where the Kassel-Exeter tasks were used again in Norway. The Hanseatic traditions[1] were common to the countries we speak about and being in Hamburg we are reminded of all the Hanseatic cities in the Nordic and Baltic countries that were linked to German centres of commerce and communication. German textbooks were used in Swedish University studies in mathematics. For example, in the 1960s, books by Knopp, *Funktionentheorie* and *Non-Euclidian Geometry,* were regularly studied in higher education courses. The book *Algebra* by van der Waerden was also part of the doctoral study curriculum. Visits from German researchers also took place. For example, Professor Doktor Emil Artin und Frau Braun (this is how they were introduced to us: Artin with all his titles and Hel Braun just as "Frau" although she was also a mathematician; Hel Braun was the very first and only lady in mathematics that I as a young doctoral student had the opportunity to meet) visited The Mathematical Society in Lund in the 1960s. These are just a few examples, there are many more.

How and why are we doing empirical research differently?

There exist some important differences in the conditions for research in didactics of mathematics. For example, the first professorships in Didactics of Mathematics were created 1992–1993 in Denmark, Finland and Norway and in Sweden in 2001 (at Luleå University of Technology), in contrast to the 60 professorships created in Germany in the 1960s. The first chair in Sweden was held for some years by R. Sträßer from

---

[1] From Wikipedia: "The Hanseatic League ... was a commercial and defensive confederation of merchant guilds and market towns in Northwestern and Central Europe. Growing from a few North German towns in the late 1100s, the league came to dominate Baltic maritime trade for three centuries along the coast of Northern Europe. It stretched from the Baltic to the North Sea and inland during the Late Middle Ages and declined slowly after 1450."

Germany. The academization of teacher education took place around 1960 but not until in the 1980s was it an explicit demand that teacher education should be research based. A scientific society of Didactics of Mathematics was created in Sweden in 1998 (called Svensk Förening för Matematik Didaktisk Forskning—SMDF) and some years earlier in Denmark and Finland. Norway and Iceland still do not formally have such societies. Thus, it can be argued that the Nordic countries were about 20–30 years behind in the development of Didactics of Mathematics as an area of research studies compared to Germany.

What was it that triggered the development in the Nordic countries? The First Mathematics and Science Study, FIMSS, created a huge debate in Sweden on this topic and the government set up committee for mathematics in school. They published their report in 1986, entitled "Matematik i skolan" [Mathematics in School], and it suggested academic courses in Didactics of Mathematics, positions and revision of the teacher education. A new teacher education started in 1988, where Didactics of Mathematics was introduced and the education became clearly research based. Student teachers were required to carry out a small research study and write a scientifically oriented report, the so called 'examensarbete'. The work called for supervision and research literature.

The International trends also swept over the Nordic countries. First in the 1960s there was a focus on the modern mathematics, followed by a back to basics movement in the 1980s, and more recently the use of ICT and problem solving. The results from TIMSS and PISA are influencing the politicians much and creating public debate in society about school mathematics. Nordic teachers were taught methods of teaching which links to Erich Wittman's view of Didactics of Mathematics as a design science. It was not seen as mere research.

In Sweden, the National Center for Mathematics Teaching (NCM), was created in the end of the 1990s and a similar centre was founded in Trondheim during 2002, the Norwegian Centre for Mathematics Education (NSMO). Again, politicians emphasized the teaching, and research was not included in their agenda. But in Norway a great effort was also given to creating research in Didactics of Mathematics. At the University of Agder (UiA) a master's education started in Didactics of Mathematics in 1994 followed by a doctoral education in 2002. Four professors of Didactics of Mathematics were hired (all women) and asked to build up a research environment and establish doctoral education. This was the first time in the Nordic countries when a group of professors could work together in Didactics of Mathematics at the same university. One of many guest researchers in the mathematics education research group, MERGA, at UiA was S. Rezat from University of Paderborn in Germany.

A huge five-year grant was given to UiA in order to set up a Nordic Graduate School in Mathematics Education, called NoGSME, which started in the beginning of 2004. Most of the Nordic universities with Didactics of Mathematics-students were linked to this Graduate School and it held about 90 doctoral students and 100 supervisors, also from the Baltic countries. Several German scholars were invited to this Graduate School to lecture, participate in summer schools or hold seminars for supervisors and thus the link to Germany was kept alive. The Graduate School was in action between 2004 and 2010 and its network for research on mathematics textbooks continued to be funded from the Nordic Research Academy until 2016 (Grevholm, 2017). In the network for research on textbooks two of our German colleagues were very active during a period of ten years. R Sträßer and S. Rezat created the link to German research on textbooks (Rezat & Strässer, 2013). NoGSME was followed by a fruitful networking collaboration institutionalised in the Nordic Society for Research in Mathematics Education (NoRME), between the Nordic and Baltic countries in the form of common conferences, the Nordic Conferences on Mathematics Education (NORMA-conferences), a common scientific journal, the Nordic Studies in Mathematics Education (NOMAD; started in 1993 and revived in 2004), and joint activities in doctoral education including courses, supervisors' seminars, summer schools in which German scholars took part.

Thus, in the Nordic countries empirical research was most often carried out by single researchers on their own, often with a lack of funding, and it resulted in a fragmented picture of results from the research. The only places where we find a larger group of researchers is at University of Agder in Norway and Umeå University in Sweden. There was no common plan for the empirical studies that were carried out in the Nordic area. The early German studies on Stoffdidaktik and textbooks had no counterpart in the Nordic Universities, and research centres such as the one in Bielefeld did not exist.

The German impact in Didactics of Mathematics in the Nordic countries

In what parts of Didactics of Mathematics can we trace the German influences from collaboration and research?

A few examples of such research areas are mentioned below.

The use of ICT in mathematics teaching and learning

The use of Information and Communication Technology (ICT) and other technological resources in mathematics teaching and learning has been a research interest in Didactics of Mathematics since the 1980s and was another link between Germany and the Nordic area. The works by Dahland (1993, 1998) at Gothenburg University and Hedrén (1990) at Linköping University illustrate studies in early days and the influence from German researchers. Dahland utilises theories in Didactics of Mathematics from the German traditions and discusses them in his dissertation. Dahland and Hedrén had many followers interested in the use of ICT, such as T. Lingefjärd, B. Grevholm, L. Engström, A-B. Fuglestad, C. Bergsten, M. Blomhöj and others. German influence was provided by Blankertz (1987), and later from Berger (1998), R. Sträßer and others.

Students' mathematical learning difficulties and dyscalculia

As mentioned above, publications on dyscalculia were already in progress from the beginning of the 1920s and continue to be of interest, especially in relation to effective pedagogies. The Swedish pedagogue Olof Magne (1967, 1998) was inspired by Henschen and several other German researchers (for example G. Schmitz, F. Padberg), when he studied pupils with difficulties in mathematics and dyscalculia. In his turn, he evoked the interest from other Nordic researchers for learning difficulties in mathematics (e.g., D. Neumann, G. Malmer, C. Ohlin, A. Engström, L. Häggblom, K. Linnanmäki).

Studies on gender and mathematics

The Swedish network on Gender and mathematics (Kvinnor och matematik) was created in 1990 after inspiration from a study group on Women and mathematics at ICME-6 in Budapest and other German colleagues who were active in Hungary. The International Organisation of Women and Mathematics, IOWME, is an ICMI affiliated group. Active members were, for example, Christine Keitel, Erika Schildkamp-Kundiger, Cornelia Niederdrenk-Felgner, Gabriele Kaiser, and later Christine Knipping. The network Women and Mathematics had members from all the Nordic countries. Some German colleagues gave presentations on issues of equity in the conferences of this society (for example Christine Keitel in 1993 and 1996, Kristina Reiss 1993 and Gabriele Kaiser on several occasions).

Teachers' and students' view of mathematics

In 1995, E. Pehkonen in Helsinki and G. Törner in Duisburg started a series of conferences with the theme Current State of Research on Mathematical Beliefs. At the beginning, it was mainly Finnish researchers such as E. Pehkonen, M. Hannula and German researchers such as G. Törner, G. Graumann, P. Berger, B. Rösken who contributed. Later the conferences became more internationally oriented and took place in many different countries. The presentations concern mathematical beliefs, attitudes, emotions and in general views of mathematics.

Problem solving in mathematics education

Another group of Finnish and German researchers in 1999 initiated the ProMath-group, also lead by Erkki Pehkonen. The aim of the ProMath group is to study and examine those mathematical-didactical questions which arise through research on the implementation of open problem solving in school. The group organizes yearly international conferences and publishes proceedings from them. The proceedings from the conferences mirror the recent international development in problem-solving, such as for example use of problem fields and open problems.

Another series of Nordic and Baltic conferences are the NORMA-conferences, started in 1994. They rotate among the five Nordic countries and there is a tradition of German participation (e.g. Graumann, 1995; Steinbring, 2005), thereby contributing with theory and methods from the German Didactics of Mathematics. The conferences held by SMDF ("Matematikdidaktiska forskningsseminarium", called the MADIF-conferences) have also opened a window to German research projects and tendencies. The contributions are often closely linked to teaching and classroom work. German teaching projects like SINUS and DISUM (Didaktische Interventionsformen für einen selbständigkeitsorientierten aufgabengesteuerten Unterricht am Beispiel Mathematik) and use of modelling can serve as examples (Blum & Leiss, 2007; Borromeo Ferri, 2007).

How can German research contribute in the future?

A wish for the future is to forge stronger bonds between Germany and the Nordic countries in Didactics of Mathematics and to develop collaboration in many ways.

Doctoral student exchange could for example be one excellent way to do this. Another could be academic teacher exchange. Guest researchers and visiting scholars are important features. German researchers are welcomed to continue to participate in the Nordic activities like conferences and courses. They are encouraged to publish in NOMAD and create joint research studies with colleagues from the Nordic countries. Such collaboration will be fruitful for all who take part.

5.5.2   Perspectives on Collaborative Empirical Research in Germany and in Poland

(By Edyta Nowinska)

Do Germany and Poland have some joint historical roots in the development of Didactics of Mathematics? Do researchers in Didactics of Mathematics in both countries collaborate on empirical research? What are the perspective on collaborative empirical research in Germany and in Poland in the future?

I have been asked these questions many times by my German colleagues interested in learning the past and the current development in Didactics of Mathematics in Poland. The geographical location of both countries might lead to the presumption of close relations between German and Polish researchers in Didactics of Mathematics. A closer look on the institutional context of their work shows, however, that such relations do not really have the form of a close cooperation, and it also gives some explanations for this fact. In the following, I give a short overview of some historical developments in Didactics of Mathematics in Poland and on the institutional context in which the most Polish researchers in Didactics of Mathematics work. Afterwards I use this background information to explain perspectives on collaborative empirical research in Germany and in Poland.

The overview given by Rudolf Sträßer on the development of German speaking didactics of mathematics points to names of German institutes and researchers who played an essential role in this development. Some of them were known in Poland due to international contacts of the Polish Professor Anna Zofia Krygowska—probably the most important person who contributed to the development of Didactics of Mathematics in Poland at the end of 1950s. She had contacts to Hans-Georg Steiner and to the Bauersfeld group from the Bielefeld *Institut für Didaktik der Mathematik*. As many other researchers in Germany, Krygowska was engaged at this time in establishing Didactics of Mathematics as a scientific discipline. In 1958 the Methods of Teaching Mathematics Department (later called the Department of Didactics of Mathematics) was created within the Faculty of Mathematics and Physics of the Pedagogical University in Cracow and Anna Zofia Krygowska was appointed to a professorship in this department.

In 1982, Krygowska succeeded in establishing the first Polish journal publishing work dealing with didactics of mathematics—Dydaktyka Matematyki ('Didactics of Mathematics') issued as the Fifth Series of *Roczniki Polskiego Towarzystwa Matem-*

atycznego ('Annals of the Polish Mathematical Society'). Since foreign journals publishing such works were not available in Poland, one important goal of the new Polish journal was publishing of English, French and German language articles dealing with Didactics of Mathematics (translated into Polish). The international work of Anna Zofia Krygowska and in her contacts to the German speaking researchers did not have the form of a collaborative empirical research, but the international discourse was important for Krygowska in searching for new ideas for the developing scientific discipline and in making her own ideas for teaching and learning mathematics more precise.

When reflecting on the current perspectives on collaborative empirical research in Germany and in Poland, one has to be conscious about the research tradition in Didactics of Mathematics initiated by Krygowska and about the institutional context of Didactics of Mathematics in Poland.

Stefan Turnau, one of Krygowska's successor describes Krygowska's work as follows: "She always thought that school mathematics should be genuine mathematics, whatever the teaching level. She also praised logical rigour, which view she embodied in her rigorous geometry textbooks. But on the other hand she was prudent enough not to allow Bourbakism to prevail in the curriculum." (1988, pp. 421–422).

Krygowska tried to create a theoretical and methodological base for the new branch of knowledge. On the core of her research were always mathematical concepts and ideas. Her research was characterized by a strong epistemic component—by attention to mathematical meanings and mathematical understandings specific to particular concepts. Her focus was mainly on the didactics of *mathematics* and less *psychology* or *sociology* of teaching and learning mathematics. The research field established by Krygowska was quite homogeneous and this seems to be a specific characteristic of research in Didactics of Mathematics in Poland to date and is also reflected in recent publications written in English. They concentrate on conceptions related to learning functions, limits, proofs, on algebraic thinking, generalization and elementary geometry or vocational education (the dominant method is thereby the case study). This seems to be a quite stable and typical characteristic of research in Didactics of Mathematics in Poland.

This characteristic can be understood if one considers the institutional context of work of Polish researchers in Didactics of Mathematics. Most of them work in institutes for mathematics and their work is evaluated (and measured in points based amongst others on publications in high ranking journals) due to the same criteria as the work of mathematicians working in these institutes. According to these criteria, publications in proceedings issued after international conferences in mathematics education—which are important for initiating and maintaining a collaborative work with other researchers in Didactics of Mathematics—have no value and therefore it is quite difficult to get financial support for the participation in such conferences (the institutional context of some of its consequences for research in Didactics of Mathematics in Poland were discussed in the panel session during the Ninth Congress of the European Society for Research in Mathematics Education in Prague; see the comments written by Marta Pytlak in Jaworski et al. (2015, p. 24). This is the first reason why conducting research related to psychological or social perspectives on teaching and learning mathematics in schools may be very problematic in this context. The second reason, even more important, is the fact that Didactics of Mathematics does not still have the status of a scientific discipline in Poland, despite of the previously mentioned tremendous work of prof. Krygowska and her successors like Ewa Swoboda and the whole Polish society of Didactics of Mathematics researchers. Consequently there is also no well-established tradition in funding empirical research on the quality of teaching and learning of mathematics in schools or on in-service teacher professional development programs or on video-based classroom research. This kind of research does not really match neither the criteria of mathematical nor pedagogical research. Thus, the institutional context has a strong influence on the work of Polish researchers in mathematics. It excludes some research questions which are essentially important for investigating and improving teaching and learning mathematics as being not valuable in the institutional context.

Many individual Polish researchers in Didactics of Mathematics use their private contacts to German researchers to learn and discuss new trends in the development of Didactics of Mathematics. The German conception of mathematics education as a 'design science' (Wittmann, 1992) and the construct of 'substantial learning environments' have been intensively used in some researchers' groups in Poland (see Jagoda, 2009), in particular in geometry. One can say that the German Didactics of Mathematics often inspires and enriches the work of individual Polish researchers but this has not been a part of a collaboration as far. If the institutional criteria for evaluating the work of researchers in Didactics of Mathematics in Poland and for funding research projects in Didactics of Mathematics do not change, such a collaboration seems to be possible only for passion and has to cope with difficulties in recognizing its value in the institutional context.

5.5.3   *Didaktik der Mathematik and Didaktika Matematiky*

(by Naďa Vondrová)

The aim of this part is to show concrete examples of how Czech *didaktika matematiky* have been influenced by German *mathematik didaktik* (or better, by individual German researchers).

The connection between the Czech (or previously Czechoslovak) and German didactics must be divided into two periods; prior to the Velvet Revolution in 1989 and after it.

Prior to 1989, the landscape of Czechoslovak teaching of mathematics was similar to that of the German Democratic Republic (see Sect. 5.2.4). There was very little connection with Western Germany, mainly because of few opportunities to travel or to have an access to international journals and books. Only a limited number of people was allowed to travel abroad and attend conferences such as ICME. Many of them were from the Faculty of Mathematics and Physics, Charles University (e.g., Jaroslav Šedivý, Oldřich Odvárko, Leoš Boček, Jiří Mikulčák and Jan Vyšín).

Jan Vyšín was also a member of Wissenschaftlicher Beirat ZDM and both he and Oldřich Odvárko were members of the editorial board of Zentralblatt für Didaktik der Mathematik. On the other hand, some German researchers such as Hans- Georg Steiner or Roland Stowasser came to visit the Faculty of Mathematics and Physics and held lectures there. I must also mention Hans-Georg Steiner's personal efforts to help researchers from Czechoslovakia, which resulted in the authorities allowing the organisation of the International Symposium on Research and Development in Mathematics Education in Bratislava in 1988 (for the proceedings see Steiner & Hejný, 1988). The event opened the door for Czechoslovak researchers to western research and new cooperation.

Western influences became more prominent in former Czechoslovakia at the time of New Math movement. Czech Mathematicians and mathematics educators closely followed the New Math movement abroad, mostly in East Germany but also in Western countries. Some articles were published in Czech journals by Miloš Jelínek and others about New Math abroad which influenced the efforts in former Czechoslovakia.

After the Velvet Revolution, completely new opportunities arose for Czech researchers. In terms of German influence, I must mention conferences like the Tagung für Didaktik der Mathematik organised by the German Gesellschaft für Didaktik der Mathematik (GDM), where Czech researchers such as František Kuřina, Milan Koman, Leoš Boček, Oldřich Odvárko, Marie Tichá, Alena Hošpesová and others were invited to present their work. Many of them were invited to give lectures at German universities (e.g. at the Mathematikdidaktisches Kolloquium at TU Dortmund). This was often accompanied by financial support through Deutsche Forschungsgemeinschaft which especially in the first years of a new regime was an important help.

The empirical turn to everyday classrooms mentioned in Sect. 5.2.3 appeared in the Czech Republic as well under the influence of Western (also German) research (see Sect. 5.2.3). Stoffdidaktik (see Sect. 5.2.1) has always been close to the Czech conception of mathematics education (from which mathematics never disappeared), and nowadays, Stoffdidaktik enlarged (see Sect. 5.3.1) taking into account the history and epistemology of mathematics and fundamental ideas of mathematics can also be found in the work of people such as Ladislav Kvasz.

From among German researchers who markedly influenced research in Czech mathematics education (as documented by publications of Czech authors), I have to name Erich Wittmann and the project Mathe 2000. The idea of mathematics as a design science resonates in Czech research (for example, in the work by František Kuřina), as well as the idea of substantial learning environments. It is often cited, for example, by Alena Hošpesová, Marie Tichá, Naďa Vondrová, and mainly Milan Hejný for whom „the design of learning environments indeed is the defining 'kernel' of research in mathematics education" (Hejný, 2012; Hošpesová et al., 2010; Stehlíková, Hejný, & Jirotková, 2011).

Other German researchers whose work is well known in the Czech Republic and used in Czech research include E. Cohors-Fresenborg, K. Hasemann, H. Meissner, G. Muller, P. Scherer, H. Steinbring, B. Wollring, E. Glasserfeld or H. Freudenthal (who had German roots). Obviously, I only mention researchers whose influence in the former Czechoslovakia can clearly be seen in publications and presentations of researchers. Surely, there were others whose work influenced research in mathematics education for individuals and who cannot be listed here.

Finally, I would like to mention common projects of Czech and German researchers which have had an impact on mathematics education both in research and in teaching in the Czech Republic. For example:

Understanding of mathematics classroom culture in different countries (Czech Republic, Germany, Italy; M. Tichá, A. Hošpesová, P. Scherer, H. Steinbring and others): The goal of the project was to improve the quality of continuous in-service education of primary school teachers. The role of a qualified joint reflection which was heavily conceptualised by the German colleagues was stressed in this process.

Motivation via *Natural Differentiation in Mathematics* (Czech Republic, Germany, Netherlands, Poland; G. Krauthausen, P. Scherer, M. Tichá, A. Hošpesová): Wittmann's concept of substantial learning environments was in the centre of the project. The team members developed the idea of a 'new' kind of differentiation starting in the first school years which is expected to contribute to a deeper understanding of what constitutes mathematics learning, by considering the learners' individual personalities and the advantage of learning in groups, as opposed to minimizing the individual differences among the pupils (see Hošpesová et al., 2010).

Implementation of Innovative Approaches to the Teaching of Mathematics (Czech Republic, United Kingdom, Germany, Greece; M. Hejný, D. Jirotková, N. Vondrová, B. Wollring, B. Spindeler and others): The project aimed to promote constructivist teaching approaches in mathematics, to change the role of the teacher in the classroom and to make the pupils more responsible for their learning. The team members developed tasks and trialled them with cooperating teachers (see Hejný et al., 2006). The main contribution of the German colleagues lied in designing elementary mathematics tasks which connected geometry, arithmetic and relations to the real world of pupils within a learning environment of regular polygons.

Communicating Own Strategies in Primary Mathematics Education (Czech Republic, Germany, the United Kingdom; M. Hejný, J. Kratochvílová, B. Wollring, A. Peter-Koop): The project focused on the design of classroom materials to structure working environments which prepare student teachers as well as classroom teachers to motivate, moderate, record and analyse classroom communication about shape, number and structure (see Cockburn, 2007). German colleagues' unique contribution was in their focus on scientific analyses of pupils' artefacts by "student teachers as researchers".

Acknowledgements Special thanks go to the editor of this book and the series editor for their helpful and constructive comments on earlier versions of this text. The same holds for Vince Geiger, who additionally turned my German English into understandable Australian English. Mistakes, inconsistencies and errors, which may still prevail, are my responsibility.

Appendix

Table of contents of an ICME-13 book on „German-speaking Traditions in Mathematics Education Research"
Editors: Hans Niels Jahnke, Lisa Hefendehl-Hebeker.
New York: Springer 2019

References

Athen, H. (Ed.). (1977). *Proceedings of the 3rd International Congress on Mathematical Education*, Aug 16–21, 1976. Karlsruhe: ICME.

Behnke, H., & Steiner, H.-G. (Eds.). (1967). *Mathematischer Unterricht für die Sechzehn- bis Einundzwanzigjährige Jugend in der Bundesrepublik Deutschland* (International Mathematische Unterrichtskommission (IMUK), Ed.). Göttingen: Vandenhoek & Rupprecht.

Berger, P. (1998). How computers affect affectivity. Role specific computer beliefs of German teachers. In T. Breiteig & G. Brekke (Eds.), *Theory into practice in mathematics education. Proceedings of Norma98, the second Nordic conference on Mathematic Education* (pp. 87–94). Kristiansand, Norway: Høgskolen i Agder.

Biehler, R., & Peter-Koop, A. (2007). Hans-Georg Steiner: A life dedicated to the development of didactics of mathematics as a scientific discipline. *ZDM Mathematics Education, 39*(1), 3–30.

Bikner-Ahsbahs, A., & Prediger, S. (Eds.). (2014). *Networking of theories as a research practice in mathematics education*. Switzerland: Springer International.

Blankertz, H. (1987). *Didaktikens teorier och modeller*. Stockholm: HLS förlag.

Blömeke, S., Hsieh, F.-J., Kaiser, G., & Schmidt, W. H. (Eds.). (2014). *International perspectives on teacher knowledge, beliefs and opportunities to learn. Teds-M Results, Advances in Mathematics Education*. Dordrecht: Springer Science+Business Media.

Blömeke, S., Bremerich-Vos, A., Kaiser, G., Nold, G., Haudeck, H., Keßler, J.-U., et al. (Eds.). (2013). *Weitere Ergebnisse zur Deutsch-, Englisch- und Mathematiklehrerausbildung aus TEDS-LT*. Münster: Waxmann.

Blömeke, S., Kaiser, G., & Lehmann, R. (Eds.). (2010a). *TEDS-M 2008 - Professionelle Kompetenz und Lerngelegenheiten angehender Primarstufenlehrkräfte im internationalen Vergleich*. Münster: Waxmann.

Blömeke, S., Kaiser, G., & Lehmann, R. (Eds.). (2010b). *TEDS-M 2008 - Professionelle Kompetenz und Lerngelegenheiten angehender angehender Mathematiklehrkräfte für die Sekundarstufe I im internationalen Vergleich*. Münster: Waxmann.

Blum, W., & Leiss, D. (2007). Investigating quality mathematics teaching: The DISUM-project. In C. Bergsten & B. Grevholm (Eds.), *Developing and researching quality in mathematics teaching and learning. Proceedings of MADIF5*. Linköping: SMDF.

Borromeo Ferri, R. (2007). The teacher's way of handling modelling problems in the classroom. In C. Bergsten & B. Grevholm (Eds.), *Developing and researching quality in mathematics teaching and learning. Proceedings of MADIF5*. Linköping: SMDF.

Bruder, R. (2003). Vergleich der Grundlegenden Konzeptionen und Arbeitsweisen der Methodik des Mathematikunterrichts in der DDR mit denen der Didaktik der Mathematik in der BRD. In H. Henning & P. Bender (Eds.), *Didaktik der Mathematik in den Alten Bundesländern - Methodik des Mathematikunterrichts in der DDR: Bericht über eine Doppeltagung zur gemeinsamen Aufarbeitung einer getrennten Geschichte* (pp. 168–174). Magdeburg, Paderborn: Otto-von-Guericke-Universität Magdeburg, Universität Paderborn.

Burscheid, H.-J., Struve, H., & Walther, G. (1992). A survey of research. *Zentralblatt für Didaktik der Mathematik, 24*(7), 296–302.

Cockburn, A. D. (Ed.). (2007). *Mathematical Understanding 5-11. A practical guide to creative communication in Mathematics*. London: Paul Chapman Publishing.

Dahland, G. (1993). *Datorstöd i matematikundervisningen*. Rapport 1993:08. Institutionen för pedagogik. Göteborg: Göteborgs universitet.

Dahland, G. (1998). *Matematikundervisningen i 1990-talets gymnasieskola*. Rapport 1998:05. Institutionen för pedagogik. Göteborg: Göteborgs universitet.

Drenckhahn, F. (Ed.). (1958). *Der Mathematische Unterricht für die Sechs- bis Fünfzehnjährigen Jugend in der Bundesrepublik Deutschland*. Göttingen: Vandenhoek & Rupprecht.

Graumann, G. (1995). The general education of children in mathematics lessons. In E. Pehkonen (Ed.), *NORMA-94 Conference*. Helsinki: Department of Teacher Education, University of Helsinki.

Grevholm, B. (Ed.). (2017). *Mathematics textbooks, their content, use and influences*. Oslo: Cappelen Damm Akademisk.

Hedrén, R. (1990). *Logoprogrammering på mellanstadiet* [Programming in Logo at intermediate level of compulsory school]. Linköping Studies in Education 28 (Doctoral thesis). Linköping: Linköping University.

Hejný, M. (2012). Exploring the cognitive dimension of teaching mathematics through scheme-oriented approach to education. *Orbis Scholae, 6*(2), 41–55.

Hejný, M., et al. (2006). *Creative teaching in mathematics*. Prague: Faculty of Education, Charles University (in Prague).

Henning, H., & Bender, P. (Eds.). (2003). *Didaktik der Mathematik in den alten Bundesländern - Methodik des Mathematikunterrichts in der DDR. Bericht über eine Doppelta-*

gung zur gemeinsamen Aufarbeitung einer getrennten Geschichte. Magdeburg, Paderborn: Otto-von-Guericke-Universität Magdeburg, Universität Paderborn. https://www.tu-chemnitz.de/mathematik/geschichte/lehrerausbildung.pdf.

Henschen, S. E. (1920). *Klinische und Anatomische Beiträge zur Pathologie des Gehirns. Teil 5: Über Aphasie, Amusie und Akalkulie* [Papers on Brain Pathology from Clinical Research and Anatomy, part 5 on aphasia, and acalculia; title translated by RS]. Stockholm: Nordiska.

Hošpesová, A., et al. (2010). *Ideas for Natural Differentiation in primary mathematics classrooms. Geometrical environment.* Rzeszów: Wydawnictwo Univwersytetu Rzeszowskiego.

Jagoda, E. (2009). Substantial Learning Environment "Tiles". *Didactica Mathematicae, Annales Societatis Mathematicae Polonae, 32*(5), 121–151.

Jahner, H. (1978/1985). *Methodik des Mathematischen Unterrichts* (Vol. 5). Heidelberg, Germany: Quelle and Meyer.

Jahnke, H. N., & Hefendehl-Hebeker, L. (Eds.). (2019). *Traditions in German-speaking mathematics education research.* New York: Springer.

Jaworski, B., Bartolini Bussi, M. G., Prediger, S., & Nowinska, E. (2015). Cultural contexts for European research and design practices in Mathematics Education. In K. Krainer & N. Vondrová (Eds.), *CERME9. Proceedings of the Ninth Congress of the European Society for Research in Mathematics Education* (pp. 7–33). Prague: Charles University, ERME.

Kaiser, G., Blömeke, S., König, J., Busse, A., Döhrmann, M., & Hoth, J. (2017). Professional competencies of (prospective) mathematics teachers—Cognitive versus situated approaches. *Educational Studies in Mathematics, 94*(2), 161–182.

Katz, D. (1913). *Psychologie und Mathematischer Unterricht.* Leipzig: Teubner.

Keitel, C. (1995). Beyond the numbers game. In B. Grevholm & G. Hanna (Eds.), *Gender and mathematics, an ICMI study.* Lund: Lund University Press.

Krauss, S., et al. (Eds.). (2017). *FALKO: Fachspezifische Lehrerkompetenzen. Konzeption von Professionswissenstests in den Fächern Deutsch, Englisch, Latein, Physik, Musik, Evangelische Religion und Pädagogik.* Münster: Waxmann.

Kunter, M., Baumert, J., Blum, W., et al. (Eds.). (2013a). *Cognitive activation in the mathematics classroom and professional competence of teachers. Results from the Coactiv Project. Mathematics Teacher Education* (Vol. 8). New York: Springer Science+Business Media.

Kunter, M., Klusmann, U., Baumert, J., Richter, D., Voss, T., & Hachfeld, A. (2013b). Professional competence of teachers: Effects on instructional quality and student development. *Journal of Educational Psychology, 105*(3), 805–820.

Kuřina, F. (2016). *Matematika jako pedagogický problém. Mé didaktické krédo* [Mathematics as a pedagogical problem. My didactic credo]. Hradec Králové: Gaudeamus.

Kvasz, L. (2015). Über die Konstitution der symbolischen Sprache der Mathematik. In G. Kadunz (Ed.), *Semiotische Perspektiven auf das Lernen von Mathematik* (pp. 51–67). Berlin: Springer Spektrum.

Lietzmann, K. J. W. (1916). Methodik des Mathematischen Unterrichts. In J. Norrenberg (Ed.), *Handbuch des Naturwissenschaftlichen und Mathematischen Unterrichts* (Vol. 3). Leipzig: Quelle & Meyer.

Lietzmann, K. J. W. (1951). *Methodik des Mathematischen Unterrichts.* Heidelberg: Quelle & Meyer.

Magne, O. (1967). *Matematiksvårigheter hos barn i åldern 7-13 år.* Stockholm: SLT-förlag.

Magne, O. (1998). *Att lyckas med matematik i grundskolan.* Lund: Studentlitteratur.

Matematik i skolan. (1986). *Översyn av undervisningen i matematik inom skolväsendet* [Overview of the teaching in the school system]. Stockholm: Liber.

Oehl, W. (1962). *Der Rechenunterricht in der Grundschule* (1st ed.). Hannover: Hermann Schroedel.

Pollak, H. O. (1979). The interaction between mathematics and other school subjects. In International Commission on Mathematical Instruction (Eds.), *New trends in mathematics teaching* (pp. 232–248). Paris: UNESCO.

Rezat, S., & Strässer, R. (2013). Methodologies in Nordic research on mathematics textbooks. In B. Grevholm, P. S. Hundeland, K. Juter, K. Kislenko, & P-E. Persson (Eds.), *Nordic research in didactics of mathematics—Past, present and future.* Oslo: Cappelen Damm Akademisk.

Schubring, G. (2012). 'Experimental Pedagogy' in Germany, Elaborated for mathematics a case study in searching the roots of Pme. *Research in Mathematics Education, 14*(3), 221–235.

Schubring, G. (2014). Mathematics education in Germany (modern times). In A. Karp & G. Schubring (Eds.), *Handbook on the history of mathematics education* (pp. 241–256). New York: Springer.

Schubring, G. (2016). Die Entwicklung der Mathematikdidaktik in Deutschland. *Mathematische Semesterberichte, 63*(1), 3–18.

Shulman, L. S. (1986). Those who understand: Knowledge growth in teaching. *Educational Researcher, 15*(2), 4–14.

Shulman, L. S. (1987). Knowledge and teaching: Foundations of the new reform. *Harvard Educational Review, 57*(1), 1–23.

Stehlíková, N., Hejný, M., & Jirotková, D. (2011). Didactical environments Stepping and Staircase. In M. Pytlak, T. Rowland, & E. Swoboda (Eds.), *Proceedings of CERME7* (pp. 2909–2911). Poland: University of Rzeszow.

Steinbring, H. (2005). Mathematical symbols: Means of describing of constructing reality. In C. Bergsten & B. Grevholm (Eds.), *Conceptions of mathematics.* Linköping: SMDF.

Steiner, H.-G. (1978). Didaktik der Mathematik - Einleitung. In H.-G. Steiner (Ed.), *Didaktik der Mathematik* (pp. 9–48). Darmstadt: Wissenschaftliche Buchgesellschaft.

Steiner, H.-G., & Hejný, M. (Eds.). (1988). *Proceedings of the international symposium on research and development in mathematics education.* Bratislava: Komenského Univerzita.

Streitlien, Å., Wiik, L., & Brekke, G. (2001). *Tankar om matematikkfaget hos elever og lærarar* [*Thoughts about the subject mathematics in students and teachers*]. Oslo: Læringssenteret.

Strunz, K. (1962). *Pädagogische Psychologie des Mathematischen Denkens* (Vol. 4). Heidelberg: Quelle & Meyer.

Tatto, M. T. (Ed.). (2013). *The Teacher Education and Development Study in Mathematics (TEDS-M) policy, practice, and readiness to teach primary and secondary mathematics in 17 Countries.* Technical Report. Amsterdam: IEA.

Tatto, M. T., et al. (2012). *Policy, practice, and readiness to teach primary and secondary mathematics in 17 countries: Findings from the IEA Teacher Education and Development Study in Mathematics (TEDS-M).* Amsterdam: IEA.

Tiedemann, K. (2017). Beschreibungen im Prozess: Eine Fallstudie zur fachbezogenen Sprachentwicklung im Kontext unterschiedlicher Darstellungen. In D. Leiss, M. Hagena, A. Neumann, & K. Schwippert (Eds.), *Mathematik und Sprache: Empirischer Forschungsstand und unterrichtliche Herausforderungen* (pp. 63–79). Münster: Waxmann.

Turnau, S. (1988). In Memoriam: Anna Sophia Krygowska. *Educational Studies in Mathematics, 19*(4), 420–422.

Voigt, J. (1996). Empirische Unterrichtsforschung in der Mathematikdidaktik. In G. Kadunz, H. Kautschitsch, G. Ossimitz, & E. Schneider (Eds.), *Trends und Perspektiven: Beiträge zum 7. Symposium zu „Didaktik der Mathematik"* in Klagenfurt vom 26. - 30.09.1994 (pp. 383–389). Wien: Hölder-Pichler-Tempski.

Winter, H. (1995). Mathematikunterricht und Allgemeinbildung. *Mitteilungen der Gesellschaft für Didaktik der Mathematik, 61,* 564–597.

Wittmann, E. C. (1992). Mathematikdidaktik als "design science". *Journal für Mathematik-Didaktik, 13*(1), 55–70. Enlarged English version: Wittmann, E. C. (1995). Mathematics Education as a Design Science. *Educational Studies in Mathematics, 29*(3), 355–374.

Chapter 6
Didactics of Mathematics as a Research Field in Scandinavia

Frode Rønning

Abstract This chapter presents an overview of the development of didactics of mathematics as a research domain in the three Scandinavian countries, Denmark, Norway and Sweden. This presentation is linked to the development of the school system and teacher education. Some important trends that have been of particular importance to each of the countries will be described and an account of the current situation will be given. At the end there is a section about collaborative projects that have taken place in the whole Nordic and Baltic area.

Keywords Competencies · Critical mathematics education · Didactics of mathematics · Mathematical modelling · Mathematics and democracy · Scandinavia

6.1 Introduction

In this chapter I will present some important aspects of didactics of mathematics as a research field in the three countries of Denmark, Norway and Sweden. The term *didactics of mathematics* corresponds to the terms used in the national languages about the research field: *matematikdidaktik* in Danish and Swedish and *matematikkdidaktikk* in Norwegian. These terms correspond to mathematics education in English. I will use the term mathematics education when I refer to the enterprise where teaching and learning of mathematics takes place. It will not be possible to cover the topic in full within the scope of the chapter, so what is presented will not be an exhaustive account of all activity. I have based my presentation on available literature, existing webpages, information that colleagues have kindly provided for me upon request and my own knowledge of the situation.

I will refer to Denmark, Norway and Sweden as the Scandinavian countries. These three countries share many similarities, both in the ways the societies are organised and the values that are shared in areas such as democracy, equality and equity. In

F. Rønning (✉)
Department of Mathematical Sciences, Norwegian University
of Science and Technology, 7491 Trondheim, Norway
e-mail: frode.ronning@ntnu.no

© The Author(s) 2019   153
W. Blum et al. (eds.), *European Traditions
in Didactics of Mathematics*, ICME-13 Monographs,
https://doi.org/10.1007/978-3-030-05514-1_6

the way the school systems and teacher education are organised one can also find many similarities between the three countries. Since didactics of mathematics is linked closely to both school system and teacher education, it is natural that the research field also shares many features across the countries. An aspect of some relevance is also that the languages in the three countries are very similar and can in general be used across the borders without major difficulties. Collaboration in the region very often also includes Iceland and Finland, which, together with Norway, Sweden and Denmark form the Nordic countries. In recent years, this collaboration has increasingly also included the Baltic states (Estonia, Latvia and Lithuania). My main presentation will not include Finland, Iceland and the Baltic states. However, there are important events and institutions where it is obvious that they should be included. The final section of the chapter is dedicated to an account of these.

After the Introduction (Sect. 6.1), I will present some historical background (Sect. 6.2) and then give a presentation of the early development of didactics of mathematics as a research field in Scandinavia (Sect. 6.3). Sections 6.4, 6.5 and 6.6 will present further development and more recent activity in the countries Denmark, Norway and Sweden, respectively (presented in alphabetical order). Finally, Sect. 6.7 is devoted to common initiatives covering the whole Nordic and Baltic region.

Earlier attempts to give an overview of research in the Nordic region have been made, for example, by Björkqvist (2003a), who made a list of research problems in Nordic research on didactics of mathematics. The items from this list that are most relevant for the Scandinavian countries are presented below.

- Research on realistic problem solving in mathematics
- Research on problem solving as an example of school mathematics as situated practice
- Research on connecting the teaching of mathematics in elementary schools to everyday experiences
- Research on the role of mathematics education in society and the political dimensions of mathematics education
- Research on the effects of the hand-held calculators in mathematics education in elementary school
- Research with a strong emphasis on understanding children with special difficulties in mathematics
- Research in the phenomenographic paradigm
- Research on mathematical misconceptions and common errors in school mathematics
- Research on strategies of students solving mathematical problems
- Research on mathematical operativity
- Research on women and mathematics (Björkqvist, 2003a).

In another report, where Björkqvist was commissioned to do a survey of research in the field in Sweden (Björkqvist, 2003b), he mentions some areas that he finds to be particularly prominent in Sweden. Two of these areas are phenomenographic research and research on mathematics and democracy. Mathematics and democracy is also an important field in the two other Scandinavian countries. I will discuss these and some other areas in some detail later. Björkqvist's reports may have presented a

Table 6.1 Research paradigms

	ESM	JRME	NOMAD
Studies of learning and cognition, including problem-solving strategies	10	6	5
Studies of outcomes of interventions, including teaching approaches and experiments	5	8	2
Suggesting and implementing theoretical/analytic constructs or frameworks	7	1	3
Uncovering beliefs, attitudes, affects or identities with teachers and students	6	0	4

Table 6.2 Methodological approaches

	ESM	JRME	NOMAD
Conceptual and theoretical investigations	10	1	3
Qualitative empirical investigations	22	10	7
Quantitative empirical investigations	6	6	7

fair description of the activity up to 2003, but, as will become clear later, the activity has increased tremendously, so a similar list 15 years later would have become much longer and more diverse. Also, some of the areas on Björkqvist's list may not be so important anymore.

It is hard to identify any dominating theories or paradigms in Scandinavian research, although some commonalities can be seen. Mogens Niss attempted to do a quantification of preferred research study paradigms and choices of methods in didactics of mathematics to see whether the Nordic countries stood out in any way from the rest of the world (Niss, 2013). He picked all papers published from April 2011 to April 2012 from three journals, *Educational Studies in Mathematics (ESM)*, *Journal for Research in Mathematics Education (JRME)* and *Nordic Studies in Mathematics Education (NOMAD)*,[1] and classified each paper as belonging to one of 10 research paradigms and using one of three methodological approaches. The total number of papers came to 72: 38 in *ESM* and 17 each in *JRME* and *NOMAD*. In the *ESM* and *JRME* papers, three Nordic authors were involved and in the *NOMAD* papers one non-Nordic author was involved. The results from this classification might give an idea about which paradigms or methods are particularly dominant in the Nordic area.[2] Table 6.1 shows the results on research paradigms presented by Niss, including only the four paradigms that were represented with 10 or more papers. Table 6.2 shows the identified methodological approaches.

---

[1] A journal for the Nordic and Baltic region (see Sect. 6.7).

[2] Niss did not distinguish between Scandinavia and the rest of the Nordic/Baltic region.

That the number of papers in *ESM* was a little more than twice the number of papers in each of the two other journals indicates that, for example, the topic of beliefs and attitudes is overrepresented in *NOMAD* compared to the two other journals and that intervention studies is somewhat underrepresented. Also, conceptual and theoretical investigations are clearly underrepresented. The overrepresentation in beliefs and attitudes can be explained by the fact that *NOMAD* published a thematic issue on belief research during this period (Vol. 16, No. 1–2, June 2011). This issue contained five articles, four of which Niss categorised under this research paradigm. It is also worth mentioning that belief research has for a long time had a particularly strong position in Finland. Researchers from Finland have kept a close connection to the Mathematical Views (MAVI) group (Pehkonen, 2012), and indeed there was another thematic issue of *NOMAD* (Vol. 17, No. 3–4, December 2012) on this topic published shortly after the period covered by Niss (2013) in his survey.

The results presented in Tables 6.1 and 6.2 come from a small set of data, so one cannot draw strong conclusions from them, but they may give an indication of what kind of research has been in focus in the chosen period. Regarding the issue of nearness to mathematics, there is great variation, both between the countries and within the countries. Some of the research has come out of departments of education, where the subject may not be so much in focus and where the leading researchers have their main background in general education (pedagogy). It may seem that nearness to mathematics in the research has become stronger in recent years and that the activity in didactics at departments of mathematics has been increasing. Most of the research has been empirical and has had a close connection to the classroom and/or to teacher education. This reflects the fact that most of the research on didactics of mathematics has grown out of institutions where teacher education has been an important activity. Some of the research has been based on design of learning environments but also much of the research has had an ethnographic style, where the learning environments have been investigated 'as they are', without intervention from the researchers.

6.2 Historical Background

Didactics of mathematics as an independent scientific discipline can be said to have existed for a little more than 50 years. The late 1960s saw such events as the founding of the journal *Educational Studies in Mathematics* in 1968 and the first ICME conference in 1969. Almost at the same time, research centres for didactics of mathematics were established in several countries, such as the Institute for the Development of Mathematical Education (later the Freudenthal Institute) in the Netherlands, the Shell Centre in the UK and the Institut für Didaktik der Mathematik in Germany (Gjone, 2013, p. 183). However, it can be claimed that the field of study had already begun to develop in the late 19th century. An important event was the establishing of the first International Commission on the Teaching of Mathematics (ICMI) in 1908 at the Fourth International Congress of Mathematicians in Rome, with Felix Klein as its first president (Kilpatrick, 1992, p. 6). ICMI conducted its first survey in 1912 and, according to Kilpatrick, this survey 'reported that university lectures on mathematics education[3] (to supplement mathematics lectures) were being offered in the United States, Great Britain, Germany and Belgium' (1992, p. 5).

In the history of ICMI, Denmark plays a special role among the Scandinavian countries. Poul Heegaard, who had studied with Klein in Göttingen, was a delegate to the first ICMI in 1908 and also wrote the report from Denmark to be included in the 1912 survey. Heegaard became one of three Vice Presidents of ICMI in 1932. At that time he had left Denmark for a professorship at the University of Oslo, which he held from 1918 until his retirement in 1941 (Furinghetti & Giacardi, 2012). Later, Svend Bundgaard and Bent Christiansen had important roles in ICMI. Bundgaard was a member of the Executive Committee from 1963 to 1966 and Christiansen served three terms (12 years) as Vice President from 1975. Also, in the years after Christiansen stepped down, the Danish representation in central bodies of ICMI continued to be strong. Mogens Niss became member of the Executive Committee in 1987 and later served as Secretary General for two terms, 1991–1998. Shortly afterwards, a somewhat urgent situation came up in connection with the organising of the 10th ICME conference. It had been assumed that ICME-10 in 2004 should be held in Brazil, but this turned out not to be possible, so an alternative host country had to be found. It was then decided that ICME-10 should be arranged as a collaboration between Denmark, Sweden, Norway, Iceland and Finland. Copenhagen was chosen as the venue, and Mogens Niss became Chair of the International Programme Committee for ICME-10 (Hodgson & Niss, 2018). Hosting ICME-10 may be seen as the peak of a very long-standing connection between Denmark and ICMI.

In the Scandinavian countries in the early 20th century there was a rising interest in issues concerning the teaching of mathematics[4] at school level, including some empirical research. In 1919, K. G. Jonsson obtained his doctoral degree at Uppsala University, Sweden, with a study based on interviews with pupils where he observed and categorised their ways of thinking when solving arithmetic problems (see Johansson, 1986). Bergsten (2002, p. 35) also mentions Karl Petter Nordlund as a pioneer from Sweden. He published his book *A Guide for the First Teaching of Arithmetic (Vägledning vid den första undervisningen i räkning)* in 1910. The growing interest in educational issues in the early 1900s can be seen in connection with the development of the school system. In the mid-1800s, a state school system for all had been established as a principle in the Scandinavian countries, which led to a need for educating teachers. Teacher education developed in similar ways in all three countries, with a distinction between educating teachers for compulsory school (known as *folkeskole* in Norwegian and Danish, *folkskola* in Swedish) and for post-compulsory school (*gymnas/ium* in Norwegian/Danish, *läroverk* in Swedish). Institutions educating teachers for compulsory school (Grades 1–7) were referred to as seminaries (*seminar/ier* in the national languages) and educated general teachers who were expected to teach all school subjects. The system in Sweden differed

---

[3]It is not clear which terms were used to describe these lectures in the various countries and languages.
[4]The term *didaktik(k)* was not yet in use in the Scandinavian languages.

somewhat from the system in Norway and Denmark in the sense that in Sweden there were special seminaries educating teachers for Grades 1 and 2. Later, when compulsory school was extended from seven to nine years, Sweden introduced differentiated education for the lower, middle and upper grades (Linné, 2010). In Norway and Denmark, however, the idea of a teacher education covering all of compulsory school was maintained much longer. In Norway, different study programmes for Grade 1–7 teachers and Grade 5–10 teachers[5] were introduced as late as in 2010.

Initially, becoming a teacher was an option only open to men, but gradually women were also admitted. In Sweden this happened in 1859 (Linné, 2010), with some seminaries exclusively for women then being established. Some of these female teachers played important roles in communicating modern ideas in the teaching and learning of mathematics, ideas that were far ahead of the common ways of thinking at the time. One example is Anna Kruse from Sweden, who published her book *Visual mathematics (Åskådningsmatematik)* in 1910. This book was meant as a guidebook for other teachers. In the preface to the first edition, she writes that

> it is not only a question of giving the child knowledge in a school subject but it is a matter of providing the child a means for acquiring one of the most important factors for the future profession—whatever it may be—clear logical thinking, ability to make judgments, and a practical view. (Kruse, 1910/2010, pp. 29–30, my translation)

It is a fundamental principle in the Scandinavian countries that schooling is part of a democratic endeavour. In the Norwegian context, Telhaug and Mediaas (2003) connect this feature of the school system to the emerging trend in the 19th century that all citizens have a responsibility to take part in the development of the society. These ideas can be traced in didactics of mathematics up to present day through the field of mathematics and democracy, which I will return to later. All three Scandinavian countries were at this time closely connected, as Norway had been part of Denmark for several hundred years until 1814 and then became part of a joint kingdom with Sweden until 1905.

In the first half of the 20th century, compulsory schooling lasted for seven years. A big change in the school system came in the 1950s and early 1960s with an extension from seven to nine years. This also indicated some changes for the subject of mathematics. In the old system, the subject in compulsory school was called *regning/räkning* (arithmetic), but in the new system the term *matematik/k* (mathematics) came into use, including in the lower grades. At the same time, also the so-called New Math[6] started to appear on the international scene, and the ideas from New Math also spread to the Scandinavian countries. A very important event in the introduction of New Math ideas in the mathematics curriculum was the seminar held at Royaumont in France in 1959 (OEEC, 1961). Here, arguments were advanced to let the ideas that had been developed in mathematics as a science over the last 50 years, for instance, through the work of the Bourbaki group, also permeate school mathematics. The outcome of this was, as is well known, a school mathematics with strong emphasis on logic and set theory.

---

[5] Since 1997, Norway has had 10 years of compulsory schooling, starting at the age of six.
[6] *Moderne matematikk* in Norwegian. *Ny matematik* in Danish and Swedish.

After the Royaumont seminar, a Nordic committee for the modernisation of mathematics education was established, with four members each from Denmark, Finland, Norway and Sweden. This committee was active from 1960 to 1967. In the recommendations from the committee (Nordisk udredningsserie, 1967) one can find the ideas from the Royaumont seminar expressed in terms of specific topics that were necessary to learn, for instance, set theory. The argument for this is that it will convey to the pupils an understanding for basic concepts. 'Elementary concepts and symbols from set theory make it possible to present the material simpler and clearer'. However, one is also aware of the danger of introducing symbols: 'On the other hand, one has to be careful so that symbols are not experienced as abstract and difficult to understand' (Nordisk udredningsserie, 1967, p. 173, my translation). New Math is often associated with logic and set theory, but it also introduced the concept of function in school mathematics, a concept that has kept its place while many of the formal structures from logic and set theory have been abandoned (Prytz & Karlberg, 2016).

The ideas of the New Math can be traced back to Felix Klein's Erlanger Programm (Klein, 1872). Through the connection between Felix Klein and Poul Heegaard, Klein's ideas had a strong influence in Denmark in the early 1900s. It seems that the New Math movement became stronger in Denmark and Sweden than in Norway. In Norway, New Math textbooks were used in some schools but they did not dominate the school system to the same extent as in Sweden and Denmark. In the Norwegian National Curriculum from 1974, concepts from set theory and logic are referred to as 'support concepts' (*hjelpebegreper*) (Kirke- og undervisningsdepartementet, 1974, pp. 143–144). The development in Denmark can, in addition to the early influence by Poul Heegaard, be explained by the work of Svend Bundgaard and Bent Christiansen. Bundgaard, a professor of mathematics at Aarhus University from 1954, was very influential regarding both university mathematics and school mathematics in Denmark. Bent Christiansen had a strong influence on the National Curriculum in Denmark, promoting New Math ideas, and he wrote several textbooks, both for schools and for teacher education. At the upper secondary (*gymnasium*) level, New Math ideas are clearly expressed in the textbook system *Textbook for the New Upper Secondary School* (*Lærebog for det nye gymnasium*) by Erik Kristensen and Ole Rindung that began in 1962. Rindung was also involved in writing the National Curriculum in Denmark (OEEC, 1961; Rønn, 1986). His central position in Denmark is also shown by the fact that he participated at the Royaumont seminar and that he was also a member of the Nordic Committee. Also Erik Kristensen and Bent Christiansen, as well as Agnete Bundgaard, who wrote textbooks for primary school, were members of the Nordic Committee (Nordisk udredningsserie, 1967, p. 220). It is my impression that the members of the Nordic Committee from the other two countries did not to the same extent influence the development of the subject in their respective countries.

6.3   Didactics of Mathematics Emerging as a Research Discipline in Scandinavia

Despite some examples of educational research involving mathematics from the early years of the 20th century, one cannot speak of didactics of mathematics as an independent research discipline in the Scandinavian countries until the 1970s. The early research was done by general educators (pedagogues), using mathematics classrooms as a frame for their empirical studies, and this is still happening. Teachers of mathematics at the teacher training colleges (*lærerhøgskoler/seminarier*) were not expected to do research and usually neither did they have the qualifications to do it. At the universities, teacher training (for lower and upper secondary school) involved doing a short add-on after taking a degree based on studies in the subject itself. Each university usually employed at most one person to take care of the actual teacher training part of mathematics teacher education. There were mathematicians at the universities with strong interest and engagement in educational matters, and some of them played important roles in developing teacher education and mathematics as a school subject although they did not do research in didactics of mathematics. The content of the teacher training in the school subjects was mostly focused on technical matters: *how* to carry out the teaching. The more theoretical parts were dealt with in the subject of pedagogy. In the 1970s there was a turn towards including topics about how learning takes place and such topics as motivation for learning in individual subjects, with the subject of mathematics being quite far ahead in this development. It was also around this time that the expression *matematik(k)didaktik(k)* came into use. Earlier, at least in Norway, one had talked about *metodikk* (methods, i.e., methods for teaching) instead of *didaktikk*. A book written by Solvang and Mellin-Olsen (1978) for use in teacher education in Norway can serve as an example of the transition from methods to didactics. The title of the book is *Mathematics Subject Methods* (*Matematikk fagmetodikk*), which points to the aspect of methods, but the content is wider and more modern in that it also contains learning theories and theories about different rationales for learning mathematics. In Sweden, Christer Bergsten pinpoints the use of the term *didaktik* to the mid-1980s and claims that the first course with the title *matematikdidaktik* was given at Linköping University in 1985 (Bergsten, 2002, p. 37). Personally, I remember that when I did my teacher training in 1982, the title of the course was *matematikkdidaktikk*, and the main literature was the book *The Psychology of Learning Mathematics* (Skemp, 1971), emphasising constructivist/cognitivist learning theories.

   In the early years, the most important institution for research in didactics of mathe-matics in Denmark, perhaps in Scandinavia as a whole, was Danmarks Lærerhøjskole (The Royal Danish School of Educational Studies), an institution where teachers could get continuing education. The institution later changed its name to Denmark Pedagogical University and is today part of Aarhus University, but still located in Copenhagen. In the 1970s, new universities were established in Denmark: Roskilde University in 1972 and Aalborg University in 1974. These universities were meant to be different from the 'old' Danish universities in Copenhagen and Aarhus, with a profile based on problem-based and project-organised work in groups. From around 1980, Roskilde University emerged to become a centre for didactics of mathematics in Denmark, and this university also engaged in educating mathematics teachers with a strong didactical component for upper secondary school, which made it rather exceptional in Denmark. Until recently, Roskilde has probably had the highest activity in didactics of mathematics in Denmark. At Aalborg, didactics of mathematics also developed as a research field, with a strong profile towards socio-political aspects of mathematics education. The result of the activity at Danmarks Lærerhøjskole was that Denmark was far ahead of the other two countries in the early years.

In Norway, teachers at the teacher training colleges (*lærerhøgskoler*) gradually started to become acquainted with research literature in didactics of mathematics. Some of them also started doing research, and a few went abroad to get a Ph.D. in didactics of mathematics. Around 1970, a new type of higher education institution was developed in Norway known as regional colleges (*distriktshøgskoler*), and some of these developed into mini-universities, offering traditional university subjects such as mathematics. This development is somewhat similar to the development in Denmark that resulted in the universities in Roskilde and Aalborg. Some of the regional colleges were established at places where there already was a teacher training college nearby, in some cases even in the same town. In Kristiansand, Kristiansand lærerhøgskole and Agder distriktshøskole coexisted until they merged into Agder University College (Høgskolen i Agder), now University of Agder, in 1994. This merger brought together a strong group of mathematicians and a strong group of teacher educators in mathematics. In the same year as the merger took place, a two-year master's programme (*hovedfag*) in didactics of mathematics was established. This developed further into a Ph.D. programme, starting in 2002. In the development that took place at Agder, Trygve Breiteig, one of the senior teacher educators and didacticians in Norway, played a very active role. The programme at Agder is still the only Ph.D. programme in Norway dedicated specifically to didactics of mathematics, although it is possible to do a Ph.D. in the field at other universities, either in a programme based in pedagogy (education) or in mathematics. The Ph.D. programme at Agder led to the appointment of professors from abroad, and this programme has been instrumental in providing Norwegian universities with Ph.D. graduates in didactics of mathematics. In Norway, all teacher education has recently been made into five-year master's programmes. This puts certain formal requirements on the scientific qualifications of the staff, which has created a situation where the demand is much higher than the supply. The development in teacher education over a number of years has led to a strong increase in the activity in didactics of mathematics in Norway.

Regarding the situation in Sweden, the activity has grown considerably from 1980 onwards. At least 20 dissertations concerning learning of and education in mathematics were defended in the period 1981–1999 (e.g., Strässer, 2005). Most of the dissertations have been submitted within educational sciences but with a focus on mathematics, and among the Swedish universities, the Gothenburg University stands out as being responsible for a large number of the dissertations.

An important event influencing the situation in Sweden happened in March 2000, when the board of the Bank of Sweden Tercentenary Foundation (Riksbankens Jubileumsfond) offered to support didactics of mathematics in Sweden by setting up a national graduate school in mathematics with a profile towards didactics of mathematics (*matematik med ämnesdidaktisk inriktning*). This will later be referred to as the Swedish Graduate School. All mathematics and general education departments in Sweden were invited to participate in the graduate school, with 10 departments chosen and 20 students admitted when the activity started in 2001 (Leder, Brandell, & Grevholm, 2004, p. 170). The Swedish Graduate School had the following effect: In addition to increasing the research activity in didactics of mathematics in Sweden, many departments of mathematics have become more engaged in didactical issues and the activity has spread to more universities and university colleges (*högskolor*).

The University of Umeå holds a special position when it comes to didactics of mathematics in Sweden. At Umeå, plans for a Ph.D. programme in didactics of mathematics were made in the mid-1990s, before the Swedish Graduate School became a reality. The activity at Umeå and the Swedish Graduate School helped Sweden make a leap forward among the Scandinavian countries regarding research in didactics of mathematics.

After this exposition of the early development, in the next three sections I will go into more detail about the development in each of the three countries. In these sections I will describe some of the areas that have been researched in each of the countries up to the present time.

6.4 The Development in Denmark

I have shown that in the beginning Denmark was far ahead of the other Scandinavian countries in establishing didactics of mathematics as a research area, starting at Danmarks Lærerhøjskole and continuing in particular at the universities in Roskilde and Aalborg. Today the situation has to a large extent turned around. In the period from 1988 to 2010, a number of external funding opportunities enabled Danish universities to educate Ph.D. candidates. However, when the funding ended, the activity was greatly reduced and only a few of the Ph.D. candidates were able to find university positions. The development in Sweden resulting from the Ph.D. programme at Umeå and the Swedish Graduate School and in Norway with the Ph.D. programme at Agder seem to have had more lasting value. In addition, teacher education both in Sweden and Norway has become increasingly more research based, contributing to a strong demand for research-qualified staff. In Denmark, teacher education for compulsory school is still to a large extent carried out by people without research qualifications, and the didactical component of teacher education for upper secondary school is rather small. Structural changes among and within the universities have also made didactics of mathematics less visible. Danmarks Lærerhøjskole, which had been an independent university, is now reduced to a department within Aarhus University. At Roskilde, the university has decided to close down both mathematics as an independent subject of study and the teacher education programme. This programme was rather unique in the Danish context as the teachers it educated for upper secondary school were given good backgrounds in didactics of mathematics. In recent years, several senior researchers have retired and not been replaced or have moved to other countries. There are currently some small research groups in didactics of mathematics at the universities and a few active researchers at the university colleges. I have for some years been involved as an external evaluator of proposals for Ph.D. scholarships in Denmark. Based on experiences from the most recent evaluation process, it seems that the recruitment of Ph.D. students in didactics of mathematics is slightly improving. It is also clear that the activity in the field at Copenhagen University is increasing. This I will also return to later. In the following sections I will go into some detail about fields that are and have been of particular importance in Denmark.

6.4.1   Mathematical Modelling

From the beginning, the programme in mathematics at Roskilde University was based on ideas from mathematical modelling, with Mogens Niss as the central researcher, followed by Morten Blomhøj. One aspect of how Niss sees mathematical modelling is 'to *perform active modelling* in given contexts, i.e., mathematising and applying it to situations beyond mathematics itself' (Niss & Højgaard, 2011, p. 58). Mathematical modelling in the Danish tradition is very much connected to handling problems from real life. This leads to the formulation of a task and further to systematisation, mathematisation, mathematical analysis, interpretation/evaluation and finally to validation by going back to the real-life situation. This is illustrated in the modelling cycle in Fig. 6.1.

Fig. 6.1 The modelling cycle (Blomhøj & Kjeldsen, 2006, p. 166, with permission)

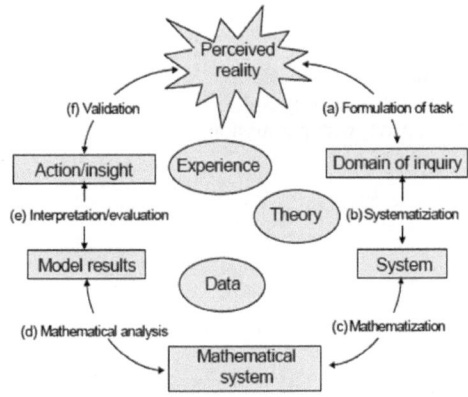

In the literature, mathematical modelling is used with slightly different connotations. Note that the Danish tradition is close to applications of mathematics to real life problems, whereas modelling is used in a wider context, for example, in the Dutch tradition, in the concept of emergent modelling (e.g., Gravemeijer, 2007).

6.4.2 Mathematical Competencies

In 2000, the Danish Ministry of Education established a group, led by Mogens Niss, for the project Competencies and Mathematical Learning (Kompetencer og matematiklæring [KOM]). The idea behind the project was to develop an alternative to the traditional idea of thinking about a subject in terms of the syllabus (*pensumtenkning*) when designing content of and goals for school subjects. Mathematics was chosen as the first subject for the new approach. This work led to the idea of describing certain competencies which, seen together, constitute what may be called *mathematical competence.* The complete report from the group was published in Danish (Niss & Jensen, 2002) and later translated into English (Niss & Højgaard, 2011). The group identified eight competencies, divided into two groups: 1. The ability to ask and answer questions in and with mathematics and 2. The ability to handle mathematical language and tools. The first group contains the following four competencies: mathematical-thinking competency, problem-tackling competency, modelling competency and reasoning competency. The second group is composed of representing competency, symbol and formalism competency, communicating competency, and aids and tools competency. The eight competencies are often presented in the 'competency flower' shown in Fig. 6.2.

This work has had great impact on mathematics education, not only in Denmark and the other Scandinavian countries but also more widely, for instance, in Germany (see Blum, Drüke-Noe, Hartung, & Köller, 2006).

Fig. 6.2 The competency flower (Niss & Højgaard, 2011, p. 51, with permission)

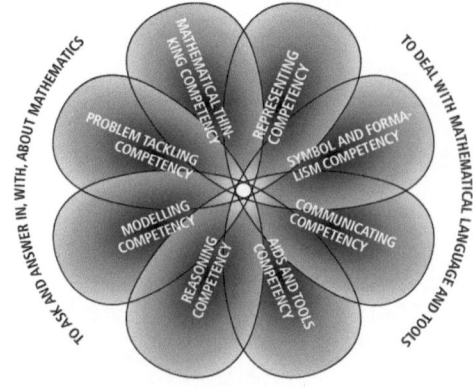

6.4.3   Political Aspects of Mathematics

Under this heading I have placed two important trends that have been particularly strong in Denmark. One of these is critical mathematics education, represented by Ole Skovsmose. This trend has also been important in Norway, which I will cover in that section. Skovsmose (2010) describes how in the early 1970s he developed an interest in critical education and then in particular in critical mathematics education. This development was connected to the student movements that swept through Europe from 1968. Out of this evolved a view that university studies should serve not only academic but also political and social interests. The most comprehensive presentation of his ideas can be found in the book *Towards a Philosophy of Critical Mathematics Education* (Skovsmose, 1994).

Also Skovsmose's concepts of landscape of investigation and exercise paradigm (Skovsmose, 2001) connect to critical mathematics education. Skovsmose contrasts the approach to teaching where the teacher and the textbook are the authorities in the classroom with that of an investigative approach. The former he denotes the exercise paradigm and the latter a landscape of investigation. In the exercise paradigm, the relevance of the matter taught is not questioned, there is one and only one correct answer and it is the role of the student to acquire the knowledge presented by the textbook and the teacher. A landscape of investigation is linked to, for example, project work, which he describes as being 'located in a "landscape" which provides resources for making investigations' (Skovsmose, 2001, p. 123). Skovsmose links this idea to critical mathematics education and the term *mathemacy*, which 'refers not only to mathematical skills, but also to a competence in interpreting and acting in a social and political situation structured by mathematics' (p. 123). In this view, mathematics is not only a subject to be learnt but also a subject to be thought about and reflected upon. Skovsmose takes the stance that there should be room for critical voices in mathematics education and he sees no possibilities for that within the exercise paradigm. In a later publication, Skovsmose (2003) recognises that his concept of the exercise paradigm is very much akin to the concept of exercise discourse (*oppgavediskursen*) that Stieg Mellin-Olsen had defined based on his empirical work with Norwegian teachers several years earlier (Mellin-Olsen, 1991). Mellin-Olsen describes the role of exercises in mathematics education as structuring the lessons. The exercises have a beginning and an end, and the end is marked by an answer that may be found in a list of solutions (*fasit*). The exercises follow in a sequence, and when one is finished, the next exercise awaits. The final goal is the exam (Mellin-Olsen, 2009, p. 2). Although there are clear similarities between the work of Mellin-Olsen and the work of Skovsmose, there are also some clear differences in the way they approach critical aspects of mathematics education. While Mellin-Olsen certainly was critical of the ways mathematics was taught in school, it can be said that Skovsmose to a larger extent also expressed critical views towards mathematics itself (Skovsmose, 2002).

The second trend that I will place under this heading is mathematics and democracy. This goes back to an initiative in the late 1980s that turned out to have important consequences for the whole Nordic area. Gunhild Nissen, a professor of education at Roskilde University, got funding for a five-year project entitled Mathematics Education and Democracy (*Matematikundervisning og demokrati*). This led to establishing a network of mathematicians, didacticians, teachers and researchers from education and psychology. The work soon led to an extension of the network outside of Denmark to include the other Nordic countries, and the first Nordic symposium on research in didactics of mathematics was held in Gilleleje, Denmark, in 1990. The project was founded on a humanistic-democratic approach to mathematics and, given the similarities between the Nordic countries, it was believed that there could be a common platform for collaboration. A central idea in the project was a desire to change the common perception of mathematics as a formalistic subject and to challenge the view that, besides basic arithmetic, people at large did not need to obtain substantial knowledge of mathematics. Nissen claims that with the more widespread use of mathematics, not only as a basis for technology, natural sciences and economy but also for administrative systems and models forming the basis for political decisions, it is necessary that everybody can take a well-informed and critical stance to the use of mathematics in order to be able to participate in a democratic society (Nissen, 1994, pp. 58–59). The project was later recognised as a Nordic project under the auspices of the Nordic Academy for Advanced Study (Nordisk Forskeruddannelsesakademi), and funding became available to provide scholarships for Ph.D. students (Nissen, 1993). One of the lasting outcomes of the work on the project is the journal *Nordic Studies in Mathematics Education* (*NOMAD*). The Mathematics Education and Democracy project gave financial support for the planning meetings for *NOMAD*, which came out with its first issue in October 1993 (Johansson, 1993). For more about *NOMAD*, see Sect. 6.7 of this chapter.

6.4.4   The French Tradition in Denmark

In the early years, there was not much activity in didactics of mathematics at Copenhagen University. However, this has changed in recent years. Now there is an active research group around Carl Winsløw. His research is to a large extent directed to research on the teaching and learning of mathematics in higher education (e.g., Gravesen, Grønbæk, & Winsløw, 2017). Theoretically, Winsløw maintains a strong connection to the French tradition, the Theory of Didactical Situations, commonly known as TDS (Brousseau, 1997) and the Anthropological Theory of the Didactic, commonly known as ATD (e.g., Bosch & Gascón, 2014). These theoretical approaches have traditionally not been well known in the Scandinavian tradition, but the work of Winsløw has contributed to a more widespread knowledge about them and also led to them being used by other Scandinavian researchers (e.g., Måsøval, 2011).

6.5 The Development in Norway

When didactics of mathematics as a scientific discipline started to emerge in Norway, the whole community of didacticians was very small, comprising around 20–25 persons in the whole country, mostly confined to teacher training colleges (*lærerhøgskoler*). Until around 1990, mathematics at teacher training colleges was a very small subject. Typically, each college would employ one or two mathematics teachers, without research qualifications or research possibilities. It is questionable whether all of these could be referred to as didacticians, but they certainly taught mathematics to prospective teachers. There was also some activity at the universities, in particular at the universities in Oslo and Bergen. Gunnar Gjone at the University of Oslo wrote a Ph.D. thesis on the New Math movement in 1983 and later worked on curriculum development and the use of ICT in mathematics, among other things. I will limit my detailed description of the early research to two persons, Gard Brekke and Stieg Mellin-Olsen. I have made this choice because these two had a strong influence on the development of didactics of mathematics in Norway and because they represent two different theoretical traditions: a constructivist tradition and a socio-cultural tradition.

6.5.1 The Constructivist Tradition

Gard Brekke (1943–2009) obtained his Ph.D. from the University of Nottingham at the Shell Centre for Mathematical Education in 1991. He played a very important role in didactics of mathematics in Norway for several decades, both as a researcher, a teacher educator (at Telemark University College) and by taking on tasks for the national authorities. Brekke, together with Gunnar Gjone, took the initiative to create the Quality in Mathematics Education (Kvalitet i matematikkundervisningen [KIM]) project in the early 1990s. This project, funded by the Norwegian Ministry of Education, aimed at developing a library of diagnostic tasks in mathematics that covered most areas of mathematics in all of compulsory school and parts of upper secondary school. The project also aimed to survey beliefs and attitudes about mathematics held by pupils in school (Brekke, 1994). The diagnostic material was intended to be used by teachers, and Brekke wrote a leaflet meant to serve as a guide to teachers for how to use the material (Brekke, 2002). Here Brekke gives his contribution to characterising knowledge in mathematics by asking the question 'what does it mean to know mathematics?' Brekke's answer to this is that mathematical competency consists of five components:

1. Factual knowledge, e.g., definitions, notation and conventions.
2. Skills, defined as well-established procedures, such as knowing how to compute the product of multi-digit numbers.
3. Conceptual structures, exemplified by multiplicative structures.

4. General strategies, defined as the ability to choose suitable skills to solve a problem from an unknown situation.

5. Attitudes. (Brekke, 2002, pp. 4–9, my translation)

The fifth component is motivated by Brekke (2002) by saying that 'our view (both as a teacher and as a pupil) on mathematics, will decide how the teacher teaches the subject, and how the pupil meets the subject matter' (p. 9, my translation). Brekke's framework for mathematical competence precedes the competency framework developed by Mogens Niss (Niss & Jensen, 2002), but it is not as detailed as this. One important difference is that unlike Niss, Brekke includes an affective component. In this sense, Brekke is in line with the framework by Kilpatrick, who includes productive disposition as one of five strands (Kilpatrick, Swafford, & Findell, 2001, p. 116).

Diagnostic teaching is based in the constructivist/cognitivist tradition and a central concept of this is the concept of a cognitive conflict. To create such a conflict, it is necessary to design tasks in particular ways so that they can uncover misconceptions. The tasks developed for the KIM project were inspired by a tradition that was very strong in the UK from the 1970s onwards. Much work on diagnostic teaching was done, in particular at the Shell Centre in Nottingham (e.g., Bell, 1993). Since Gard Brekke spent several years at the Shell Centre in the late 1980s and obtained his Ph.D. there in 1991, it is natural that this tradition influenced his work. Given Brekke's central position in the Norwegian community of teacher educators, it is also natural that it influenced didactics of mathematics in Norway more widely.

6.5.2   The Socio-cultural Tradition

A different theoretical tradition in Norway can be identified when going into the work of Stieg Mellin-Olsen (1939–1995) at the University of Bergen. In 1977 he had already published his book *Learning as a Social Process* (*Læring som sosial prosess*; Mellin-Olsen, 1977). In the first chapter of this book, Mellin-Olsen discusses Piaget's theory of knowledge. This discussion develops into a critique of Piaget, but first he extracts what he sees as the main point in Piaget's theory: '[*K*]*nowledge is connected to work*. It is the way we work that decides what kind of knowledge we develop' (Mellin-Olsen, 1977, p. 18, my translation, emphasis in original). From this he develops his concepts causal (or relational) understanding (*årsaksforståelse*) and instrumental understanding (*instrumentell forståelse*) and links these concepts to Piaget's concepts of operational and figurative knowledge (p. 20). The concepts relational understanding and instrumental understanding are usually attributed to Richard R. Skemp and are widely known from his paper with the same title (Skemp, 1976). However, Skemp and Mellin-Olsen collaborated in the 1970s, and they refer to each other's work when they write about different forms of understanding. Thus far, one may say that Mellin-Olsen is aligned with constructivism, but later in his book he presents a critique of Piaget (Mellin-Olsen, 1977). This critique is rooted in an experiment that Mellin-Olsen was involved in at a lower secondary school in Bergen that was set up with a test group using an alternative curriculum and a control group using the traditional curriculum. The experiment lasted for two years and, probably to the researchers' disappointment, the result was that the test group did not do any better than the control group. However, Mellin-Olsen and his colleagues realised that when evaluating educational experiments, the traditional psychological variables are insufficient to explain the outcome. Mellin-Olsen (1977) writes that it turned out to be necessary to also take into account anthropological and sociological considerations, such as pupils' thoughts about the school and what they expect from it (p. 22).

Mellin-Olsen brought ideas into the field of didactics of mathematics that presented an alternative to the constructivist/cognitivist tradition that was prevailing at the time. In his book published in 1987 he makes clear references to Vygotsky and his successors and he devotes the first chapter of this book to activity theory (Mellin-Olsen, 1987). However, in *Learning as a Social Process* (Mellin-Olsen, 1977) there are no references to Vygotsky. When writing the 1977 book he may not even have been familiar with Vygotsky's theories. At that time, Vygotsky's writings were not easily available outside of the Soviet Union and, if at all, only in Russian. Although the first English version, not quite complete, of *Thought and Language* came out in 1962, it was not until the late 1970s that more of Vygotsky's work became available (Kozulin, 1985, pp. *liv–lvi*). In his 1987 book, Mellin-Olsen writes that 'still only two modest books by Vygotsky are available in English' (p. 29), and in this book (e.g., p. 20) he elaborates on his critique of Piaget. Mellin-Olsen is clearly concerned with the school's and the subject of mathematics' role as reproducers of social injustice, and in this sense he can be seen as a public educator: in the terms of Ernest (2000), as one whose mathematical aims are '[e]mpowerment of learners as critical and mathematically literate citizens in society' (p. 6).

6.5.3 Further Work Within the Socio-cultural Paradigm

In the more recent activity in Norway one can also see that socio-cultural theories play an important role. At the University of Agder, much of the activity is generated from a socio-cultural standpoint. This can be seen for example in the work on inquiry-based learning initiated by Barbara Jaworski (e.g., Jaworski, 2006). Socio-cultural traditions are also visible in the research using activity theory (e.g., Jaworski & Goodchild, 2006). Also contributing to the socio-cultural tradition at Agder was Maria Luiza Cestari. She collaborated closely with Roger Säljö from Gothenburg, who held a visiting professorship at Agder for some years.

In Bergen, the tradition of Stieg Mellin-Olsen has mainly been continued at the teacher education part of what now is Western Norway University College. For many years, Marit Johnsen-Høines was the leading person at this institution. She is now retired and the group is now led by Tamsin Meaney. She has an international reputation as a researcher on multicultural and multilingual mathematics classrooms.

Since she took up the position in Bergen, this is now one of the research areas of the group, in addition to critical perspectives on mathematics education and mathematics for young children.

At the University of Tromsø, the northernmost university in Norway, there is a research group with a particular interest in mathematics related to the Sami culture in Norway. Anne Birgitte Fyhn has led or has been involved in several research projects with connection to the Sami culture (e.g., Fyhn, 2010). In this respect, there are similarities between the activity in Tromsø and Bergen, and these similarities have also led to collaboration between the two groups (e.g., Fyhn, Meaney, Nystad, & Nutti, 2017).

6.5.4 Research on University Didactics

Research in didactics of mathematics at higher education is a rapidly growing field internationally. This is noticeable through the increasing number of publications in the area, with new journals appearing, and also through centres in various countries focusing on university mathematics. This area has also gained ground in Norway, to a large extent due to the Centre for Excellence in Education, Centre for Research, Innovation and Coordination of Mathematics Teaching (MatRIC),[7] hosted by the University of Agder. MatRIC was established in 2014 and aims to support developmental projects for mathematics at higher education but also to do research through staff and Ph.D. students. There is also some activity on research on mathematics at higher education at the Norwegian University of Science and Technology (NTNU), mainly connected to projects for reforming the basic courses in mathematics and statistics for engineering students (e.g., Rønning, 2017).

6.5.5 Classroom Research and Research on Aspects of Teacher Education

In addition to the University of Agder, the University of Stavanger is another of the 'new' universities (former university colleges) in Norway that has developed a strong group of researchers in didactics of mathematics. The activity at Stavanger has been dominated by work on mathematical knowledge for teaching (MKT), as developed by Deborah Ball and colleagues (e.g., Ball, Thames, & Phelps, 2008). The group at Stavanger has worked on translating the MKT test items from English to Norwegian and testing them out on Norwegian teachers (e.g., Fauskanger, 2015). Another area that has attracted interest at Stavanger is lesson study. Raymond Bjuland was the leader of an interdisciplinary project, Teachers as Students, where lesson study as

---

[7] www.matric.no.

a method was tried out in several subjects, including mathematics (e.g., Munthe, Bjuland, & Helgevold, 2016).

NTNU, which after a recent merger now is the largest university in Norway, includes a large Department of Teacher Education with some 25–30 persons working mainly with mathematics in teacher education for compulsory school. The majority of this group came to NTNU from Sør-Trøndelag University College through the merger. Many of these people are young, often with a Ph.D. in mathematics, and are in a developing phase as researchers in didactics of mathematics. The group at the Department of Teacher Education has a close connection to the small group at the Department of Mathematical Sciences, for instance, through the collaborative project Language Development in the Mathematics Classroom. This is a project where researchers and teachers at two primary schools work together over several years to study the importance of the learning environment for young learners' development of mathematical thinking and understanding (e.g., Dahl, Klemp, & Nilssen, 2017; Rønning & Strømskag, 2017). There is also on-going research regarding various aspects of pre-service teachers' development into the teaching profession (e.g., Enge & Valenta, 2015).

6.5.6   Large-Scale Studies

Much of the research in Norway is of a qualitative character. However, Norway has actively participated in the TIMSS and PISA studies, and this activity is confined to the University of Oslo, where the research group in didactics of mathematics has been responsible for the national reports from these studies (e.g., Bergem, Kaarstein, & Nilsen, 2016; Kjærnsli & Jensen, 2016).

6.6   The Development in Sweden

Presently, Sweden is the Scandinavian country with the by far largest activity in didactics of mathematics and with the largest number of researchers in the field. Several of the candidates from the Swedish Graduate School (see Leder et al., 2004) now hold professorships at various universities in Sweden. Not only is the level of activity high but the activity is also very diverse, so it will not be possible to cover all areas in this chapter. I will start by presenting in some depth two of the 'old' areas and then proceed to some of the more current activity. The presentation of the current situation is strongly based on information that has been sent to me upon request from colleagues in Sweden.

6.6.1 Low Achievement

It seems natural in a presentation of Swedish work on didactics of mathematics to start with the work on low achievement. In this area, Olof Magne was a pioneer (Magne, 1958). Typical for Magne's studies is that they cover a large number of pupils and/or are followed up over several years. His study in 1953 covered all 6000 pupils in three random school districts of the compulsory school system in Gothenburg (Engström & Magne, 2010, pp. 335–336). In the late 1970s, Magne initiated the large project known as the Middletown Mathematics Project (*Medelsta-matematik*). Middletown, a fictitious name for a municipality of 25,000 inhabitants, was selected to represent an average municipality in Sweden. Its compulsory school (*grundskola*) had about 2000 pupils. In 1977, all pupils at the school from Grades 1 to 9 were tested using assessment material developed as part of the project. The Middletown study was repeated in 1986 and again in 2002 (see Engström & Magne, 2010 for more details).

6.6.2 Phenomenography and Variation Theory

In a paper from 1981, Ference Marton, at the Department of Education at the University of Gothenburg, presented a new approach to research which he denoted phenomenography. In this paper he makes a distinction between making statements about the world and making statements about people's ideas about, or experiences of, the world (Marton, 1981, p. 178). Initially, this research did not have anything in particular to do with mathematics. However, it turned out that it would be used in a number of research projects where mathematics played an important role. The successful application of variation theory to the study of teaching and learning of mathematics may explain the important and perhaps even dominating role of the Department of Education at Gothenburg at a rather early stage of didactical research.

After a large number of empirical studies using the phenomenographic approach, there emerged a more theoretical approach known as variation theory. Runesson and Kullberg (2010) denote this as the theoretical turn of phenomenography. The main idea here is that 'learning takes place, knowledge is born, by a change in something in the world as experienced by a person' (Marton & Booth, 1997, p. 139). A basic principle for variation theory is that if something varies and something else remains constant, it is more likely that the thing that varies will be noticed, and further it is assumed that the phenomenon that is noticed is more likely to be learned. In collaboration with researchers from Hong Kong, an approach known as learning study was developed (Runesson & Kullberg, 2010). This can be described as an intervention model where teachers and researchers work together designing lessons on a specific topic with specific learning goals.

6.6.3   Learning by Imitative and Creative Reasoning

Johan Lithner at Umeå University has developed a framework for reasoning. He argues that although there are several frameworks for describing, for example, stages of understanding (such as Hiebert & Lefevre, 1986; Sfard, 1991; Skemp, 1976), 'there are not many that aim at characterising the reasoning itself' (Lithner, 2008, pp. 255–256). In his work he has used several terms to characterise different types of reasoning, for instance, plausible reasoning and reasoning based on established experiences (Lithner, 2000). Later, he turns to the main categories, creative reasoning and imitative reasoning, with imitative reasoning split into the subcategories of memorised reasoning and algorithmic reasoning (Lithner, 2008). A main point was to characterise the key aspects of imitative reasoning, which was found to be the dominating type of reasoning in the empirical data. The framework of creative and imitative reasoning has also been used in a number of other studies and by other authors, both within and outside of the Nordic community. Another example from Umeå is by Ewa Bergqvist who studied the type of reasoning required to solve exam problems (Bergqvist, 2007). A synthesis of the research outcomes in this area can be found in Lithner (2017).

6.6.4   Assessment

There is work on assessment taking place at many universities in Sweden. At Umeå University, formative assessment is an important field, and the work there is closely connected to classroom practice and the results are meant to be used both by practitioners and as a background for educational research and school development projects, in collaboration between the university, schools and municipalities (e.g., Andersson & Palm, 2017a, b).

At Stockholm University, there is a research group known as the PRIM Group, which is focusing in particular on various aspects of assessment. The PRIM Group has a national responsibility for developing tests and various types of assessment material for use in Swedish schools, and it is also responsible for the Swedish part of the PISA project. The PRIM Group is led by Astrid Pettersson and has more than 20 people working with particular tasks, from designing tests to performing statistical analyses of results (e.g., Pettersson & Boistrup, 2010).

6.6.5   Mathematics and Language

Mathematics and language is a research area with high international activity, such as with topic groups at CERME and ICME. This area is also represented in Swedish research. The overarching purpose of the research in this area is to increase the understanding of the role language and communication has for knowledge and learning in mathematics by studying, for example, connections between types of argumentation and students' understanding of mathematical explanations. At Umeå University, Magnus Österholm, Ewa Bergqvist and Anneli Dyrvold are working in this field (e.g., Dyrvold, Bergqvist, & Österholm, 2015; Österholm & Bergqvist, 2013). Also at Uppsala University there is a research group working on mathematics and language, specialising on text analysis using linguistic methods and collaborating with researchers in linguistics (e.g., Bergvall, Folkeryd, & Liberg, 2016).

6.6.6 Early Learning of Mathematics

At Stockholm University there is a project entitled The Acquisition of Year One Students' Foundational Number Sense in Sweden and England (FoNS, i.e., foundational number sense). This is a comparative study between Sweden and England with the aim of investigating how teachers and parents in Sweden and England support year one students to learn the skills defined as FoNS. Furthermore, it is also the aim of the project to help teachers and parents in their work supporting children's learning of FoNS. The project is built on the idea that children with poorly developed number sense are likely to remain low achievers throughout their schooling. The project is led by Paul Andrews and several researchers at the department contribute to the project (e.g., Sayers, Andrews, & Boistrup, 2016).

At other universities there have also been researchers working with early learning, in school and pre-school, such as Maria Johansson at Luleå who has also collaborated with researchers from Malmö University (e.g., Johansson, Lange, Meaney, Riesbeck, & Wernberg, 2014). Lovisa Sumpter at Stockholm University has also worked with pre-school children (e.g., Sumpter & Hedefalk, 2015). Another example is Jorryt van Bommel at Karlstad and Hanna Palmér at Linnaeus University who have been working with young children's exploration of probability (Van Bommel & Palmér, 2016).

6.6.7 Inclusive Mathematics Education

I have chosen this heading to cover a variety of activities, including research on political issues in mathematics education, multilingualism and gender issues. I have previously presented critical mathematics education as an area represented by Ole Skovsmose in Denmark and Stieg Mellin-Olsen in Norway. Paola Valero continued the tradition of Skovsmose at Aalborg University in Denmark, but she is currently working at Stockholm University. Valero's current research can be described as exploring the significance of mathematics education as a field where power relations are actualized in producing subjectivities and generating inclusion/exclusion of different types of students. In Valero (2017) she shows an example of her recent work.

Under this heading I will also put the work of Eva Norén, also at Stockholm University. One of her main interests has been multilingual students' opportunities to position themselves within discourses. In one paper, she studied a Grade 1 multilingual classroom to explore how students' agency is expressed in the classroom (Norén, 2015a). She has also investigated the positioning of girls and boys in mathematics classrooms (Norén, 2015b). The theoretical basis of Norén's work is Foucault's theory of discourse.

Some of the work of Lisa Björklund Boistrup (Stockholm) could also be placed under the heading of inclusive mathematics education. In her Ph.D. thesis (Boistrup, 2010), she draws on Focault's theory on discourses but her study is also based in social semiotics. I find her work to have clear aspects of inclusion, as she states that she addresses how assessment systems and processes act to benefit or disadvantage individuals or groups (2010, p. 38).

6.6.8 Research on Particular Mathematical Topics

Bergsten (2010) has made a list of research topics in Sweden, and in this list there is nothing about research on particular mathematical topics. This situation has changed since 2010. Now there are several research groups working on particular mathematical topic areas, such as algebra. At Uppsala University there is an ongoing project with the title Towards research-based teaching of algebra, which addresses various aspects of school algebra through interviews with teachers and examination of textbooks (e.g., Bråting, Hemmi, Madej, & Röj-Lindberg, 2016). This project also has a strand on historical methods, which is a speciality of the Uppsala group (e.g., Prytz, 2018). The topic of algebra has attracted much attention in Sweden over the last 15 years or so, partly due to Swedish students' poor performance on tests such as PISA.

As mentioned earlier, there has been a long tradition of doing research in didactics of mathematics at the Faculty of Education at Gothenburg University. This research now also includes research on algebra. There is a project called VIDEOMAT, which is a comparative study involving Sweden, Norway, Finland and the US. This project is based on classroom studies and interviews with teachers and has a particular focus on early algebra and use of variables (e.g., Rystedt, Kilhamn, & Helenius, 2016).

Much of the activity in the Swedish Graduate School, established in 2000, was linked to departments of mathematics, and with mathematicians taking supervision responsibility. It is reasonable to believe that this led to an increased interest in research on specific mathematical topics. An example of a thesis coming out of the Graduate School, focusing on a specific mathematical topic, in this case the concept of function, is the work by Juter (2006) at Kristianstad University. She has continued to be interested in the concept of function and related topics (e.g., Juter, 2017).

6.6.9 Research on Teacher Education

Many of the researchers in didactics of mathematics in Sweden work at universities where teacher education is an important part of the activity. Therefore, it is also natural that this is an important area of research. One example of research on teacher education is the work done by Jeppe Skott and colleagues at Linnaeus University. Skott is Danish and worked for many years at the former Danmarks Lærerhøjskole before taking up a position at Linnaeus University in Växjö. His research on and with teachers is done under the heading The Makings of a Mathematics Teacher. This is basically about professional identities of teachers in pre-school and compulsory school. The work adopts a social perspective on learning and identity and has led to development of a framework called Patterns of Participation (e.g., Palmér, 2013; Skott, 2013, 2017).

There is also research on teacher education at Karlstad University, in particular connected to professional development of teachers. Various approaches are taken, such as focusing on pupils with special needs and developmental projects involving digitalisation, such as using social media (Van Bommel & Liljekvist, 2016). Pre-service teachers' development is also studied in a project at Kristianstad University by Kristina Juter and Catarina Wästerlid. Groups of pre-service teachers for school-years 4–6 are participating in a longitudinal study about identity development in becoming mathematics teachers. Kicki Skog at Stockholm University, taking a socio-political theoretical perspective, studies how different cultures, contexts and politics affect what is learnt, how it is learnt and what it is to become a mathematics teacher (Skog, 2014).

6.7 Important Initiatives Across the Countries

There is a long tradition of collaboration between the Scandinavian countries, and indeed between all Nordic countries. In more recent years this collaboration has also come to include the Baltic states, most notably Estonia. In this section I will briefly give an account of some of the most important arenas for collaboration in the region.

6.7.1 Nordic Studies in Mathematics Education (NOMAD)

Parallel to the development of didactics of mathematics as a research domain, a need grew for an outlet to publish results from research and developmental work. The first initiatives in this direction came with the establishing of journals for mathematics teachers and teacher educators: *Matematik* in Denmark in 1973, *Nämnaren* in Sweden in 1974, and *Tangenten* in Norway in 1990. Along with the development of these professional journals, the idea grew to also have a scientific journal for the Nordic region. Mellin-Olsen (1993) reports on an initiative by Göran Emanuelsson (editor of *Nämnaren*) in 1988 that resulted in a meeting in April 1989 where a group consisting of Gunnar Gjone and Stieg Mellin-Olsen from Norway and Göran Emanuelsson and Bengt Johansson from Sweden was established to work on the idea of creating a new journal. As a result of this work came the journal *Nordic Studies in Mathematics Education* (*NOMAD*). The first issue appeared in October 1993. As described in Sect. 6.4.3, the Mathematics Education and Democracy project was important for bringing *NOMAD* into being as it provided financial support for the planning meetings.

It was decided that articles in *NOMAD* could be written in Danish, Norwegian, Swedish or English. In the beginning, most articles were written in a Scandinavian language, but, as Bengt Johansson writes in the first issue: 'The aim is that the last issue in each volume should be in English' (Johansson, 1993, p. 6, my translation). It turned out that English soon became the preferred language for publication although there have always been articles in the Scandinavian languages. In the first two full volumes (1994 and 1995), 8 out of 21 articles are in a Scandinavian language, and in the last two volumes (2016 and 2017) only 3 out of 43 articles are in a Scandinavian language. These figures also say something about the growth of the activity in didactics of mathematics research in the Nordic area. The number of published articles per volume in *NOMAD* has roughly doubled since the early years. In addition to the general increase in didactical research, the increased publishing activity in *NOMAD* is largely due to the increasing number of Ph.D. candidates, in particular in Sweden and Norway. Each volume of *NOMAD* consists of four issues and for a number of years now the last issue of each volume has been a thematic issue, related to one particular topic of research. A recent thematic issue (No. 4, 2017) is on university mathematics, reflecting also the growing interest in this particular field.

From the beginning, the responsibility for editing *NOMAD* shifted between the Nordic countries. The first four volumes were edited in Sweden, with the editorship then shifting to Norway and later to Finland and Denmark. With the growth of electronic communication, the need for having the editors in one place diminished. The country-specific editorship started to break up in 2009 when Johan Häggström joined the Danish editors Morten Blomhøj and Paola Valero, and it continued in 2010 when I replaced Paola Valero in the team. I served on the team of editors from 2010 to 2017. Johan Häggström had been managing editor since 2004, when the National Center for Mathematics Education (NCM) in Gothenburg took over the responsibility for publishing *NOMAD*. In 2012 the group of editors was extended to five to meet the increasing inflow of papers, with Kristina Juter (Sweden), Markku Hannula (Finland) and Uffe T. Jankvist (Denmark) joining the team. The current editorial team consists of Ewa Bergqvist (Sweden), Janne Fauskanger (Norway), Markus Hähkiöniemi (Finland), Tomas Højgaard (Denmark) and Johan Häggström at NCM, who is also one of two managing editors.

6.7.2 The NORMA Conferences

Around the same time that *NOMAD* was established came the idea to organise a Nordic conference in mathematics education. Winsløw (2009, p. 1) attributes the idea of a Nordic conference mainly to Erkki Pehkonen from Finland, and the First Nordic Conference on Mathematics Education (NORMA 94) was held in Lahti, Finland, in 1994. This first conference attracted many researchers from Finland and the Baltic states but not as many from the Scandinavian countries. Going through the proceedings (Pehkonen, 1995), I can identify three papers from Sweden, two from Norway and one from Denmark. The NORMA conferences have been held regularly since, with participants both from inside and outside of the Nordic region. In particular, it has been an aim to invite plenary speakers from outside of the Nordic region in addition to speakers from within the region. In Table 6.3 I have listed the NORMA conferences held so far and the names of the plenary speakers at each conference who at the time of the conference were not affiliated with a Nordic university.

Table 6.3 List of NORMA conferences

Year	City	Country	Plenary speakers
1994	Lahti	Finland	Joop van Dormolen (NL) Barbara Jaworski (UK) Thomas J. Cooney (USA)
1998	Kristiansand	Norway	Konrad Krainer (A) Marja van den Heuvel-Panhuizen (NL) Michal Yerushalmy (IL)
2001	Kristianstad	Sweden	Maria Alessandra Mariotti (I) John Mason (UK) Heinz Steinbring (D)
2005	Trondheim	Norway	Simon Goodchild (UK) Birgit Pepin (UK)
2008	Copenhagen	Denmark	Michèle Artigue (F) Paul Drijvers (NL)
2011	Reykjavik	Iceland	Núria Planas (E) Bharath Sriraman (USA)
2014	Turku	Finland	Helen M. Doerr (USA)
2017	Stockholm	Sweden	Wim Van Dooren (B)

6.7.3 The Nordic Graduate School for Mathematics Education (NoGSME)

Based on a grant from the Nordic Academy for Advanced Study, a Nordic Graduate School in Mathematics Education was established for the period 2004–2008 with Barbro Grevholm at Agder University College (now University of Agder) as chair. This came to be known as NoGSME. NoGSME played an important role in developing didactics of mathematics as a research area in the Nordic countries. From the start, about 45 supervisors and 80 doctoral students were connected to NoGSME (Grevholm, 2005, p. 61). An important activity in NoGSME was to arrange summer or winter schools for Ph.D. students, with a range of invited international experts. In addition, NoGSME supported doctoral courses held at specific universities. Formalised education of researchers in didactics of mathematics was still in an early phase in the region and there was a need for developing supervisor competence. NoGSME recognised this need and organised a range of seminars for supervisors with invited international experts. NoGSME kept close contact with *NOMAD* and the NORMA conferences. The chair of NoGSME reported extensively on the activity in every issue of *NOMAD*, and at NORMA 05 in Trondheim a special NoGSME session was held. Editors of *NOMAD* also participated at NoGSME seminars to offer help in writing papers for a research journal.

In one of her reports in *NOMAD* Grevholm (2007) starts to discuss what she refers to as the life after NoGSME. This discussion is based on comments from the evaluation panel that conducted the mid-term evaluation of NoGSME. The main concern of the evaluation panel was how to maintain the activity for doctoral students when the funding for NoGSME ended. A need for an organisation that could take responsibility for securing funding for events such as summer schools and also take responsibility for the continuation of the NORMA conferences was identified. This led to the founding of The Nordic Society for Research in Mathematics Education (NoRME)[8] in 2008 during the NORMA conference in Copenhagen (see Grevholm, 2008), with the author as the first chair. NoRME is an umbrella organisation where the national associations for mathematics education research in the Nordic and Baltic states and *NOMAD* are the members. As could be expected, it was not possible to maintain the high activity level from NoGSME after the funding ended. However, a few summer schools have been arranged under the auspices of NoRME. In addition, *NOMAD* has arranged a seminar for Ph.D. students every year since 2012. NoRME is recognised as an international organisation in mathematics education and is given an account in Hodgson, Rogers, Lerman, and Lim-Teo (2013). The present chair of NoRME is Eva Norén at Stockholm University.

Acknowledgements I am extremely grateful to a number of people for having provided information that made it possible for me to write this chapter. I owe big thanks to the following colleagues from Sweden for supplying valuable information: Paul Andrews, Lisa Björklund Boistrup, Jorryt van Bommel, Bengt Johansson, Kristina Juter, Ulla Runesson Kempe, Cecilia Kilhamn, Hanna Palmér,

---

[8] www.norme.me.

Johan Prytz, Andreas Ryve, Jeppe Skott, Timo Tossavainen, Paola Valero and in particular Johan Lithner who has also read and commented on an early version of the manuscript. Above all I want to thank Mogens Niss, who has provided much of the information about Denmark and about the early history in general. I also thank him for valuable discussions and for having read and commented on several versions of the manuscript. I emphasise that the selection of material and the way the material is presented is entirely my own responsibility.

References

Andersson, C., & Palm, T. (2017a). Characteristics of improved formative assessment practice. *Education Inquiry, 8*(2), 104–122.

Andersson, C., & Palm, T. (2017b). The impact of formative assessment on student achievement: A study of the effects of changes to classroom practice after a comprehensive professional development programme. *Learning and Instruction, 49,* 92–102.

Ball, D. L., Thames, M. H., & Phelps, G. (2008). Content knowledge for teaching. What makes it special? *Journal of Teacher Education, 59*(5), 389–407.

Bell, A. (1993). Some experiments in diagnostic teaching. *Educational Studies in Mathematics, 24,* 115–137.

Bergem, O. K., Kaarstein, H., & Nilsen, T. (Eds.). (2016). *Vi kan lykkes i realfag. Resultater og analyse fra TIMSS 2015* [We can succeed in mathematics and science. Results and analysis from TIMSS 2015]. Oslo: Universitetsforlaget.

Bergqvist, E. (2007). Types of reasoning required in university exams in mathematics. *Journal of Mathematical Behavior, 26,* 348–370.

Bergsten, C. (2002). Faces of Swedish research in mathematics education. In C. Bergsten, B. Grevholm, & G. Dahland (Eds.), *Action and research in the mathematics classroom. Proceedings of MADIF2*, Göteborg, 26–27 January 2000 (pp. 21–36). Linköping: SMDF.

Bergsten, C. (2010). Mathematics education research in Sweden: An introduction. In B. Sriraman, et al. (Eds.), *The first sourcebook on Nordic research in mathematics education* (pp. 269–282). Charlotte, NC: Information Age Publishing.

Bergvall, I., Folkeryd, J. W., & Liberg, C. (2016). Linguistic features and their function in different mathematical content areas in TIMSS 2011. *Nordic Studies in Mathematics Education, 21*(2), 45–68.

Björkqvist, O. (2003a). A sampler of research problems in Nordic mathematics education research. In A. Tengstrand (Ed.), *Proceedings of the Nordic Pre-conference to ICME10, 2004* (pp. 13–21). Gothenburg: National Center for Mathematics Education.

Björkqvist, O. (2003b). *Matematikdidaktiken i Sverige – en lägesbeskrivning av forskningen och utvecklingsarbetet* [Didactics of mathematics in Sweden—A description of the state regarding research and development]. Stockholm: Kungliga Vetenskapsakademin.

Blomhøj, M., & Kjeldsen, T. H. (2006). Teaching mathematical modelling through project work—Experiences from an in-service course for upper secondary teachers. *ZDM—Mathematics Education, 38*(2), 163–177.

Blum, W., Drüke-Noe, C., Hartung, R., & Köller, O. (Eds.). (2006). *Bildungsstandards Mathematik: konkret*. Berlin: Cornelsen-Scriptor.

Boistrup, L. B. (2010). *Assessment discourses in mathematics classrooms: A multimodal social semiotic study* (Doctoral dissertation). Stockholm, Sweden: Stockholm University.

Bosch, M., & Gascón, J. (2014). Introduction to the anthropological theory of the didactic (ATD). In A. Bikner-Ahsbahs & S. Prediger (Eds.), *Networking of theories as a research practice in mathematics education* (pp. 67–83). Cham: Springer.

Bråting, K., Hemmi, K., Madej, L., & Röj-Lindberg, A.-S. (2016). *Towards research-based teaching of algebra—Analyzing expected student progression in the Swedish curriculum for grades 1–9.* Paper presented at the 13th International Congress on Mathematics Education, ICME-13, Hamburg, 24–31 July 2016.

Brekke, G. (1994). KIM – Kvalitet i matematikkundervisning [KIM—Quality in mathematics education]. *Tangenten, 5*(1), 4–8.

Brekke, G. (2002). *Introduksjon til diagnostisk undervisning i matematikk* [Introduction to diagnostic teaching in mathematics]. Oslo: Læringssenteret.

Brousseau, G. (1997). *The theory of didactical situations in mathematics: Didactique des mathématiques, 1970–1990* (N. Balacheff, M. Cooper, R. Sutherland, & V. Warfield, Eds. & Trans.). Dordrecht: Kluwer.

Dahl, H., Klemp, T., & Nilssen, V. (2017). Collaborative talk in mathematics—Contrasting examples from third graders. *Education 3–13. International Journal of Primary, Elementary and Early Years Education,* 1–13. https://doi.org/10.1080/03004279.2017.1336563.

Dyrvold, A., Bergqvist, E., & Österholm, M. (2015). Uncommon vocabulary in mathematical tasks in relation to solution frequency and demand of reading ability. *Nordic Studies in Mathematics Education, 20*(1), 5–31.

Enge, O., & Valenta, A. (2015). Student teachers' work on reasoning and proving. In H. Silfverberg, T. Kärki, & M. S. Hannula (Eds.), *Nordic research in mathematics education: Proceedings of NORMA14,* Turku, June 3–6, 2014 (pp. 61–70). Turku: University of Turku.

Engström, A., & Magne, O. (2010). From Henschen to Middletown mathematics: Swedish research on low achievement in mathematics. In B. Sriraman, et al. (Eds.), *The first sourcebook on Nordic research in mathematics education* (pp. 333–345). Charlotte, NC: Information Age Publishing.

Ernest, P. (2000). Why teach mathematics? In S. Bramall & J. White (Eds.), *Why learn maths?* (pp. 1–14). London: Institute of Education.

Fauskanger, J. (2015). Challenges in measuring teachers' knowledge. *Educational Studies in Mathematics, 90,* 57–73.

Furinghetti, F., & Giacardi, L. (Eds.). (2012). *The first century of the International Commission on Mathematical Instruction (1908–2008). History of ICMI.* http://www.icmihistory.unito.it.

Fyhn, A. B. (2010). Sámi culture and algebra in the curriculum. In V. Duran-Guerrier, S. Soury-Lavergne, & F. Arzarello (Eds.), *Proceedings of the Sixth Congress of the European Society for Research in Mathematics Education* (pp. 489–498). Lyon: INRP and ERME.

Fyhn, A. B., Meaney, T. J., Nystad, K., & Nutti, Y. J. (2017). How Sámi teachers' development of a teaching unit influences their self-determination. In T. Dooley & G. Gueudet (Eds.), *Proceedings of the Tenth Congress of the European Society for Research in Mathematics Education* (pp. 1481–1488). Dublin: DCU and ERME.

Gjone, G. (2013). NOMAD—Nordic studies in mathematics education. The first eight years. In B. Grevholm, P. S. Hundeland, K. Juter, K. Kislenko, & P.-E. Persson (Eds.), *Nordic research in didactics of mathematics: Past, present and future* (pp. 182–198). Oslo: Cappelen Damm Akademisk.

Gravemeijer, K. (2007). Emergent modelling as a precursor to mathematical modelling. In W. Blum, P. L. Galbraith, H.-W. Henn, & M. Niss (Eds.), *Modelling and applications in mathematics education. The 14th ICMI Study* (pp. 137–144). New York, NY: Springer.

Gravesen, K. F., Grønbæk, N., & Winsløw, C. (2017). Task design for students' work with basic theory in analysis. The cases of multidimensional differentiability and curve integrals. *International Journal of Research in Undergraduate Mathematics Education, 3*(1), 9–33.

Grevholm, B. (2005). The Nordic graduate school in mathematics education—A growing network. *Nordic Studies in Mathematics Education, 10*(2), 61–62.

Grevholm, B. (2007). The Nordic graduate school in mathematics education. *Nordic Studies in Mathematics Education, 12*(4), 79–83.

Grevholm, B. (2008). The Nordic graduate school in mathematics education—Planning for the future. *Nordic Studies in Mathematics Education, 13*(2), 93–98.

Hiebert, J., & Lefevre, P. (1986). Conceptual and procedural knowledge in mathematics: An introductory analysis. In J. Hiebert (Ed.), *Conceptual and procedural knowledge: The case of mathematics* (pp. 1–27). Hillsdale, NJ: Lawrence Erlbaum.

Hodgson, B. R., & Niss, M. (2018). ICMI 1966-2016: A double insiders' view of the latest half century of the International Commission on Mathematical Instruction. In G. Kaiser, H. Forgasz, M. Graven, A. Kuzniak, E. Simmt, & B. Xu (Eds.), *Invited lectures from the 13th International Congress on Mathematical Education* (pp. 229–247). Cham: Springer.

Hodgson, B. R., Rogers, L. F., Lerman, S., & Lim-Teo, S. K. (2013). International organisations in mathematics education. In M. A. Clements, A. J. Bishop, C. Keitel, J. Kilpatrick, & F. K. S. Leung (Eds.), *Third international handbook of mathematics education* (pp. 901–947). New York, NY: Springer.

Jaworski, B. (2006). Theory and practice in mathematics teaching development: Critical inquiry as a mode of learning in teaching. *Journal of Mathematics Teacher Education, 9,* 187–211.

Jaworski, B., & Goodchild, S. (2006). Inquiry community in an activity theory frame. In J. Novotná, H. Moraová, M. Krátká, & N. Stehliková (Eds.), *Proceedings of the 30th Conference of the International Group for the Psychology of Mathematics Education* (Vol. 3, pp. 353–360). Prague: PME.

Johansson, B. (1986). Sveriges förste forskare i matematikdidaktik [The first researcher in didactics of mathematics in Sweden]. *Nämnaren, 12*(3), 6–10.

Johansson, B. (1993). En ny nordisk forskningstidskrift [A new Nordic research journal]. *Nordic Studies in Mathematics Education, 1*(1), 4–7.

Johansson, M., Lange, T., Meaney, T., Riesbeck, E., & Wernberg, A. (2014). Young children's multimodal mathematical explanations. *ZDM—The International Journal on Mathematics Education, 46,* 895–909.

Juter, K. (2006). *Limits of functions—University students' concept development* (Doctoral dissertation). Luleå, Sweden: Luleå University of Technology.

Juter, K. (2017). University students' understandings of concept relations and preferred representations of continuity and differentiability. In T. Dooley & G. Gueudet (Eds.), *Proceedings of the Tenth Congress of the European Society for Research in Mathematics Education* (pp. 2121–2128). Dublin: DCU and ERME.

Kilpatrick, J. (1992). A history of research in mathematics education. In D. A. Grouws (Ed.), *Handbook of research on mathematics teaching and learning* (pp. 3–38). New York, NY: Simon & Schuster Macmillan.

Kilpatrick, J., Swafford, J., & Findell, B. (Eds.). (2001). *Adding it up: Helping children learn mathematics*. Washington, DC: National Academy Press.

Kirke- og undervisningsdepartementet. (1974). *Mønsterplan for grunnskolen 1974* [National Curriculum for compulsory school 1974]. Oslo: Author.

Kjærnsli, M., & Jensen, F. (Eds.). (2016). *Stø kurs. Norske elevers kompetanse i naturfag, matematikk og lesing i PISA 2015* [Steady course. Norwegian pupils' competence in mathematics, science and reading]. Oslo: Universitetsforlaget.

Klein, F. (1872). *Vergleichende Betrachtungen über neuere geometrische Forschungen*. Erlangen: Andreas Deichert. http://www.deutschestextarchiv.de/book/view/klein_geometrische_1872?p= 13.

Kozulin, A. (1985). Vygotsky in context. In L. Vygotsky (Ed.), *Thought and language* (Λ. Kozulin, Ed. & Trans.) (pp. *xi–lvi*). Cambridge, MA: The MIT Press.

Kruse, A. (2010). *Åskådningsmatematik* [Visual mathematics]. Stockholm: Nordsteds. (Original work published 1910)

Leder, G. C., Brandell, G., & Grevholm, B. (2004). The Swedish graduate school in mathematics education. *Nordic Studies in Mathematics Education, 9*(2), 165–182.

Linné, A. (2010). *Lärarutbildning i historisk belysning* [Teacher education in historic light] Resource document. Lärarnas historia [The history of teachers]. http://www.lararnashistoria.se/ article/lararutbildningens_historia.

Lithner, J. (2000). Mathematical reasoning in task solving. *Educational Studies in Mathematics,* *41,* 165–190.

Lithner, J. (2008). A research framework for creative and imitative reasoning. *Educational Studies in Mathematics, 67,* 255–276.

Lithner, J. (2017). Principles for designing mathematical tasks that enhance imitative and creative reasoning. *ZDM—Mathematics Education, 49,* 937–950.

Magne, O. (1958). *Dyskalkuli bland folkskoleelever* [Dyscalculia among primary school pupils]. Gothenburg: University of Gothenburg.

Marton, F. (1981). Phenomenography—Describing conceptions of the world around us. *Instructional Science, 10,* 177–200.

Marton, F., & Booth, S. (1997). *Learning and awareness.* Mahwah, NJ: Lawrence Erlbaum.

Måsøval, H. S. (2011). *Factors constraining students' appropriation of algebraic generality in shape patterns: A case study of didactical situations in mathematics at a university college* (Doctoral dissertation). Kristiansand, Norway: University of Agder.

Mellin-Olsen, S. (1977). *Læring som sosial prosess* [Learning as a social process]. Oslo: Gyldendal Norsk Forlag.

Mellin-Olsen, S. (1987). *The politics of mathematics education.* Dordrecht: D. Reidel Publishing Company.

Mellin-Olsen, S. (1991). *Hvordan tenker lærere om matematikkundervisning?* [How do teachers think about the teaching of mathematics?] Bergen: Bergen College of Education.

Mellin-Olsen, S. (1993). Et nordisk forskningstidsskrift for matematikk fagdidaktikk [A Nordic research journal for didactics of mathematics]. *Tangenten, 4*(2), 18.

Mellin-Olsen, S. (2009). Oppgavediskursen i matematikk [The exercise discourse in mathematics]. *Tangenten, 20*(2), 2–7.

Munthe, E., Bjuland, R., & Helgevold, N. (2016). Lesson study in field practice: A time-lagged experiment in initial teacher education in Norway. *International Journal for Lesson and Learning Studies, 5*(2), 142–154.

Niss, M. (2013). Dominant study paradigms in mathematics education research—For better and for worse. Global trends and their impact on Nordic research. In B. Grevholm, P. S. Hundeland, K. Juter, K. Kislenko, & P.-E. Persson (Eds.), *Nordic research in didactics of mathematics: Past, present and future* (pp. 395–408). Oslo: Cappelen Damm Akademisk.

Niss, M., & Højgaard, T. (2011). *Competencies and mathematical learning. Ideas and inspiration for the development of mathematics teaching and learning in Denmark.* Roskilde: Roskilde University.

Niss, M., & Jensen, T. H. (2002). *Kompetencer og matematiklæring. Ideer og inspirasjon til udvikling af matematikundervisning i Danmark.* Copenhagen: Undervisningsministeriet.

Nissen, G. (1993). Nordisk forskernetverk – Initiativet matematikundervisning og demokrati [Nordic network of researchers—the initiative mathematics education and democracy]. *Nordic Studies in Mathematics Education, 1*(1), 62–65.

Nissen, G. (1994). Matematikundervisning i en demokratisk kultur [Mathematics education in a democratic culture]. *Nordic Studies in Mathematics Education, 2*(2), 58–69.

Nordisk udredningsserie. (1967). *Nordisk skolmatematik* [Nordic school mathematics]. Stockholm: Esselte AB.

Norén, E. (2015a). Agency and positioning in a multilingual mathematics classroom. *Educational Studies in Mathematics, 89,* 167–184.

Norén, E. (2015b). Positioning of girls and boys in a primary mathematics classroom. In H. Silfverberg, T. Kärki, & M. S. Hannula (Eds.), *Nordic research in mathematics education: Proceedings of NORMA14,* Turku, June 3–6, 2014 (pp. 361–370). Turku: Finnish Research Association for Subject Didactics.

OEEC. (1961). *New thinking in school mathematics.* Paris: Author.

Österholm, M., & Bergqvist, E. (2013). What is so special about mathematical texts? Analyses of common claims in research literature and of properties of textbooks. *ZDM—The International Journal on Mathematics Education, 45*(5), 751–763.

Palmér, H. (2013). *To become—or not to become—A primary school mathematics teacher* (Doctoral dissertation). Växjö, Sweden: Linnaeus University.

Pehkonen, E. (Ed.). (1995). *NORMA-94 Conference. Proceedings of the Nordic Conference on Mathematics Teaching (NORMA-94) in Lahti 1994.* Helsinki: University of Helsinki.

Pehkonen, E. (2012). Research on mathematics beliefs. The birth and growth of the MAVI group in 1995–2012. *Nordic Studies in Mathematics Education, 17*(3–4), 7–22.

Pettersson, A., & Boistrup, L. B. (2010). National assessment in Swedish compulsory school. In B. Sriraman, et al. (Eds.), *The first sourcebook on Nordic research in mathematics education* (pp. 373–385). Charlotte, NC: Information Age Publishing.

Prytz, J. (2018). The New Math and school governance: An explanation of the decline of the New Math in Sweden. In F. Furinghetti & A. Karp (Eds.), *Researching the history of mathematics education: An international overview* (pp. 189–216). Cham: Springer.

Prytz, J., & Karlberg, M. (2016). Nordic school mathematics revisited—On the introduction and functionality of New Math. *Nordic Studies in Mathematics Education, 21*(1), 71–93.

Rønn, E. (1986). Matematikundervisningen i folkeskolen 1958–1975 [Mathematics teaching in the primary school 1958–1975]. In *Uddannelseshistorie 1986, Årbog for Selskabet for Skole- og Uddannelseshistorie* (pp. 66–94).

Rønning, F. (2017). Influence of computer-aided assessment tools on ways of working with mathematics. *Teaching Mathematics and its Applications, 36*(2), 94–107.

Rønning, F., & Strømskag, H. (2017). Entering the mathematical register through evolution of the material milieu for classification of polygons. In T. Dooley & G. Gueudet (Eds.), *Proceedings of the Tenth Congress of the European Society for Research in Mathematics Education* (pp. 1348–1355). Dublin: DCU and ERME.

Runesson, U., & Kullberg, A. (2010). Learning from variation: Differences in learners' ways of experiencing differences. In B. Sriraman, et al. (Eds.), *The first sourcebook on Nordic research in mathematics education* (pp. 299–317). Charlotte, NC: Information Age Publishing.

Rystedt, E., Kilhamn, C., & Helenius, O. (2016). Moving in and out of contexts in collaborative reasoning about equations. *Journal of Mathematical Behavior, 44,* 50–64.

Sayers, J., Andrews, P., & Boistrup, L. B. (2016). The role of conceptual subitising in the development of foundational number sense. In T. Meaney, O. Helenius, M. L. Johansson, T. Lange, & A. Wernberg (Eds.), *Mathematics education in the early years: Results from the POEM2 conference, 2014* (pp. 371–396). Cham: Springer.

Sfard, A. (1991). On the dual nature of mathematical conceptions: Reflections on processes and objects as different sides of the same coin. *Educational Studies in Mathematics, 22,* 1–36.

Skemp, R. R. (1971). *The psychology of learning mathematics.* Harmondsworth: Penguin Books.

Skemp, R. R. (1976). Relational understanding and instrumental understanding. *Mathematics Teaching, 77,* 20–26.

Skog, K. (2014). *Power, positionings and mathematics. Discursive practices in mathematics teacher education* (Doctoral dissertation). Stockholm, Sweden: Stockholm University.

Skott, J. (2013). Understanding the role of the teacher in emerging classroom practices: Searching for patterns of participation. *ZDM—The International Journal on Mathematics Education, 45*(4), 547–559.

Skott, J. (2017). Towards a participatory account of learning to teach. In A. Quortrup & M. Wiberg (Eds.), *Dealing with conceptualisations of learning—Learning between means and aims in theory and practice* (pp. 133–143). Rotterdam: Sense.

Skovsmose, O. (1994). *Towards a philosophy of critical mathematics education.* Dordrecht: Kluwer.

Skovsmose, O. (2001). Landscapes of investigation. *ZDM—Mathematics Education, 33*(4), 123–132.

Skovsmose, O. (2002). Matematiken er hverken god eller dårlig – og da slet ikke neutral [Mathematics is neither good nor bad—But far from neutral]. *Tangenten, 13*(3), 22–26.

Skovsmose, O. (2003). Undersøgelseslandskaber [Landscapes of investigation]. In O. Skovsmose & M. Blomhøj (Eds.), *Kan det virkelig passe?- om matematiklæring* [Could it really be like that?—On the learning of mathematics] (pp. 143–157). Copenhagen: L&R Uddannelse.

Skovsmose, O. (2010). Critical mathematics education: In terms of concerns. In B. Sriraman, et al. (Eds.), *The first sourcebook on Nordic research in mathematics education* (pp. 671–682). Charlotte, NC: Information Age Publishing.

Solvang, R., & Mellin-Olsen, S. (1978). *Matematikk fagmetodikk* [Mathematics subject methods]. Stabekk: NKI-forlaget.

Strässer, R. (2005). *An overview of research on teaching and learning mathematics.* Stockholm: Vetenskapsrådet.

Sumpter, L., & Hedefalk, M. (2015). Preschool children's collective mathematical reasoning during free outdoor play. *The Journal of Mathematical Behavior, 39,* 1–10.

Telhaug, A. O., & Mediaas, O. A. (2003). *Grunnskolen som nasjonsbygger. Fra statspietisme til nyliberalisme* [The role of compulsory school in building the nation. From state pietism to new liberalism]. Oslo: Abstrakt Forlag AS.

Valero, P. (2017). Mathematics for all, economic growth, and the making of the citizen-worker. In T. S. Popkewitz, J. Diaz, & C. Kirchgasler (Eds.), *A political sociology of educational knowledge: Studies of exclusions and difference* (pp. 117–132). New York, NY: Routledge.

Van Bommel, J., & Liljekvist, Y. (2016). *Teachers' informal professional development on social media and social network sites: when and what do they discuss?* Paper presented at the ERME-topic conference: Mathematics teaching, resources and teacher professional development, Humboldt-Universität, Berlin. https://hal.archives-ouvertes.fr/ETC3/public/Full_Download.pdf.

Van Bommel, J., & Palmér, H. (2016). Young children exploring probability: With focus on their documentations. *Nordic Studies in Mathematics Education, 21*(4), 95–114.

Winsløw, C. (2009). Nordic research in mathematics education: From NORMA08 to the future. In C. Winsløw (Ed.), *Nordic research in mathematics education: Proceedings from NORMA08,* Copenhagen, April 21–April 25, 2008 (pp. 1–4). Rotterdam: Sense Publishers.

Chapter 7
Czech and Slovak Research in Didactics of Mathematics

Tradition and a Glance at Present State

Jarmila Novotná, Marie Tichá and Naďa Vondrová

Abstract This chapter presents the emergence of research in didactics of mathematics in the former Czechoslovakia and gives a glimpse at its present state. It is done against the background of the history of schooling in the area and with respect to international influences such as the New Math movement. Due to a limited access to international research prior to the Velvet Revolution in 1989, Czechoslovak research developed relatively independently, yet its character was similar to that of the West. An overview of research after the Revolution is divided into four streams: development of theories, knowledge and education of teachers, classroom research, and pupils' reasoning in mathematics. Each stream is described by relevant work by Czech and Slovak researchers (with a focus on empirical research) and illustrated by publications.

Keywords Mathematics education research · New math · History of schooling · Czechoslovakia · Czech republic · Slovak republic

J. Novotná (✉) · N. Vondrová
Charles University, Prague, Czech Republic
e-mail: jarmila.novotna@pedf.cuni.cz

N. Vondrová
e-mail: nada.vondrova@pedf.cuni.cz

M. Tichá
Institute of Mathematics, Czech Academy of Sciences, Prague, Czech Republic
e-mail: ticha@math.cas.cz

W. Blum et al. (eds.), *European Traditions in Didactics of Mathematics*, ICME-13 Monographs,
https://doi.org/10.1007/978-3-030-05514-1_7

7.1   Introduction

7.1.1   Aim of the Chapter[1]

The aim of the chapter is to present roots and milestones in the origin and development of didactics of mathematics (mathematics education) as a science in the countries of the former Czechoslovakia. We base our ideas on a brief history of schooling and teacher education in the region. Next, we present how situations and events accompanying the emerging field of didactics of mathematics in other countries influenced the situation in our region.

The description of gradually developing research in mathematics education is naturally divided into two periods. The first period starts around the 1960s and ends with the Velvet Revolution[2] in 1989. The communist government at that time influenced all aspects of life, including research. Access of Czech and Slovak researchers to international research was very limited, so the field evolved in its own way. Unlike in many other fields, the work in mathematics education was little influenced by ideology. The second period spans the time after 1989 when mathematics education could develop freely and connect to international research.

The overview of research in the past 25 years or so strives to highlight the main research streams in both countries. We scrutinised publications of Czech and Slovak researchers to find their focuses and results. Some of them will only be mentioned, while others that we consider to be substantial are described in more detail. Finally, we briefly describe the main perspectives and challenges of mathematics education research as we experience them in the Czech Republic (CZ) and partially in Slovakia (SK).[3]

---

[1]The basis of the text is a chapter in a Czech book about the field didactics written by the same authors in 2015 (Vondrová, Novotná & Tichá 2015). However, it has been substantially modified and augmented to include Slovak research and new research streams and to and to provide information to an international an international rather than a national audience.

[2]The Velvet Revolution was a non-violent transition of power in what was then Czechoslovakia. Demonstrations against the one-party government of the Communist Party of Czechoslovakia combined students and dissidents. The Revolution ended 41 years of one-party rule and began the dismantling of the planned economy and conversion to a parliamentary republic (https://en.wikipedia.org/wiki/Velvet_Revolution).

[3]For the sake of brevity, we will use acronyms CZ and SK when we refer to the respective parts of the former Czechoslovakia (which formally ended in 1992) or to the two newly established countries, Czech Republic and Slovak Republic after 1992.

7.1.2   Brief History of Schooling and Teacher Education in the Region

Before looking at research in didactics of mathematics, we briefly present milestones of schooling and teacher education in the countries of the former Czechoslovakia as they highlight roots from which research not only in mathematics education has grown.

First, we must mention the personality of Jan Amos Comenius (1592–1670), who became the main representative of a socially committed pedagogy, particularly in Northern Europe. In his major work *Didactica magna* (*The Great Didactic,* 1657), he developed pedagogical principles that deeply influenced education in many European countries. According to Comenius, education should be: (a) universal, regardless of sex or financial means, and it is the state's task to ensure this; (b) realistic and ideas should at every step be grounded in reality; c) physical as well as mental and moral; and (d) practical, accompanied by action and practice. Moreover, (e) more science should be taught with the advancing age of the students, and (f) all education and knowledge should be directed to improving character and piety in the individual and order and happiness in the state (Jackson, 2011). Comenius attributed the main role to teachers. It was their task to 'provide interest and an atmosphere in the classroom in which the child will wish to learn' and 'to permit the child to observe for himself and arrange for the child to have direct experience in learning by doing' (Jackson, 2011, p. 99). Comenius's other influential texts such as *Orbis Pictus* (or *Orbis Sensualium Pictus*; *The Visible World in Pictures*, 1658) and *Schola Ludus* (*Playful School*, 1654) develop his pedagogical principles further. His ability to elaborate a genuinely pedagogical interpretation of didactics led to the establishment of pedagogy 'as a truly independent science based on criteria and principles epistemologically and gnosiologically' (Maviglia, 2016, p. 59).

Comenius's ideas continue to inspire teachers and researchers world-wide even today and they can indeed be seen in reform efforts throughout the history of CZ and SK, even though their implementation was not entirely successful. As early as 1774, six-year school attendance had already become compulsory in our region, which was accompanied by the onset of organised teacher preparation in the form of several-month courses. In 1868, an act was passed according to which primary school teachers were educated in four-year teacher education institutes (Mikulčák, 2010). It was a form of secondary education, accepting students after three years of secondary schooling, that contributed to considerable enhancing of teacher education standards. Graduation from these institutes and passing of '*maturita*' (school leaving examination) did not constitute the full qualification. Prospective teachers had to complete practical education and pass an examination in pedagogical competence to acquire a professional qualification. Already at that time, teachers tried to elevate their education to a university level. Their efforts were supported by G. A. Lindner, a professor of pedagogy at the Faculty of Arts at Charles University,[4] and others (Vališová & Kasíková, 2011). However, after the establishment of Czechoslovakia in 1918,[5] the Ministry of Education as well as general public took a negative stance towards it. Consequently, primary teachers, supported by some university teachers of pedagogy, established a private School of High Pedagogical Studies using their own resources. Its two-year study program followed the study at a teacher education institute.

The preparation of secondary teachers was different. In the second half of the 19th century, teachers already had to take courses at faculties of arts or faculties of sciences. Their study mostly consisted of subject preparation and in some years also included a year of teaching finished by a practical examination (Mikulčák, 2010). Unlike primary teachers, secondary teachers complained about a lack of preparation in psychology and pedagogy.

The free development of schools and teacher preparation was disrupted by the Second World War, the period of German pressure on schools, and by the communist coup in 1948. After that, pedagogy was subordinated to ideologisation and to the Marxist-Leninist conception of scientific communism (part of which was also the communist conception of undifferentiated education). For more than 40 years (with some breaks such as the Prague Spring in 1968), there was a strong influence of the Soviet conception.

In 1946, provisions were set down for the establishment of faculties of education, mostly under universities, which educated primary teachers, so teachers of all levels of schooling were educated at the university level. However, already in 1953, the faculties were closed and, in 1959, institutes of education were set up bearing the title of regional universities. Faculties of education were brought into existence again in 1964 and have continued until today. Until 1989, teacher education in Czechoslovakia was unified at all faculties educating teachers. Not only the curricula, the textbooks and learning texts, but also the number of lessons, time allotment and students' duties were identical. The number of contact hours and study controls was rigorously defined and for future primary teachers divided into three approximately equal parts: one third was allocated for common background studies including pedagogy, psychology and teaching practicum, and the other two thirds were equally divided between two disciplines for which students were to be qualified. Future elementary teachers have since been educated as generalist teachers with an option to specialize in areas such as art, physical education and music.

---

[4]Charles University, one of the oldest European Universities, was founded in 1348. It was modelled on the universities in Bologna and Paris. The first university in the Slovak territory, Academia Istropolitana, was established in Bratislava in 1467; unfortunately, it lasted only about 20 years.

[5]After 1918, the educational systems of both main parts of Czechoslovakia also became very close because there was a lack of qualified Slovak teachers. That is why many Czech teachers came to Slovakia to teach between 1918 and 1939.

7.2   Emergence of Didactics of Mathematics as a Science

In this section, some important milestones in the development of mathematics education (or the didactics of mathematics as it is called in CZ and SK) are described, both at an international and a national level. For the international level, only those aspects that substantially influenced the Czechoslovak (and later Czech and Slovak) research will be mentioned. By 'national perspective' we mean the Czechoslovak perspective.

7.2.1   International Perspective

The main characteristic of research in mathematics education in Czechoslovakia before 1989 was a very limited access to international research. Few researchers could travel abroad to Western countries to conferences, stay at foreign universities and have access to international proceedings and journals. Yet, some links were established. For example, there were mutual visits at universities: Researchers, mostly from the Faculty of Mathematics and Physics, Charles University, went to Germany (J. Šedivý, O. Odvárko, L. Boček, J. Mikulčák and others) and vice versa (H. G. Steiner, R. Stowasser, E. Wittmann and others). A few Czech or Slovak people were members of the scientific board or the editorial board of *Zentralblatt für Didaktik der Mathematik*.

The establishment of international organisations and conferences was indeed important for the development of mathematics education internationally and to some extent influenced the onset of research in Czechoslovakia. For example, a small number of researchers attended events such as ICMEs (the International Congresses on Mathematical Education), and reports were published in the Czechoslovak journals (e.g., Ripková & Šedivý, 1986), or meetings of the International Mathematical Olympiad (e.g., J. Vyšín and academician J. Novák). An important event for the Czechoslovak research in mathematics education was the organisation of the International Symposium on Research and Development in Mathematics Education in Bratislava in 1988 and the publication of the proceedings[6] (see Steiner & Hejný, 1988). This was a rare event allowed by the communist authorities at that time.

By far the strongest influence on mathematics education, not only in Czechoslovakia, came from the New Math movement. Its spread in the country was heavily supported by Czechoslovak mathematicians and mathematics educators. This is documented, for example, by articles published in Czech journals by M. Jelínek, J. Šedivý and J. Vyšín, which influenced the movement in Czechoslovakia (see Sect. 7.2.2).

Regardless of the limited access to international results in Czechoslovakia, some researchers and their work profoundly influenced mathematics education research in Czechoslovakia (and continue to do so).[7] First, we must mention

---

[6]The conference had two follow-up meetings.

[7]Many of the seminal books on mathematics education were available to CZ and SK researchers in Russian translations only (e.g., G. Polya: *Kak resat zadacu*).

the work by H. Freudenthal (the concept of guided reinvention, 1972, 1986, 1991). Much input came from Polish and Hungarian mathematicians and mathematics educators such as Z. Krygowská (1977), Z. Semadeni (1985), S. Turnau (1980), G. Polya (1945) and T. Varga (1976). Among others, we will mention E. Castelnuovo from Italy (New Math movement), G. Brousseau from France (1997)[8] and E. Wittmann from Germany (Project Mathe 2000; mathematics education as a design science and the idea of substantial learning environments, 1981). Obviously, we only mention researchers whose influence in the former Czechoslovakia can clearly be seen in publications and presentations of CZ and SK researchers. There were certainly others whose work influenced research in mathematics education for individuals and who cannot all be listed here.

In the next section, we will elaborate on the roots of mathematics education research from the national perspective, which comprises both countries in question.

7.2.2   National Perspective

In the first half of the 20th century, we cannot yet speak about scientific work in mathematics education. In Czechoslovakia, there was no specific institution or workplace whose task would have been to work in mathematics education, especially since it was not part of teacher education. The only books dealing with mathematics education issues were textbooks for pedagogical institutes such as K. Hruša's *Methodology of Counting* (1962). Many distinguished mathematicians were interested in education, which can be seen in their authorship of mathematics textbooks. For example, B. Bydžovský and J. Vojtěch published textbooks in the first half of the 20th century and E. Čech during the Second World War. From today's perspective, it is interesting that E. Čech valued the mathematical knowledge of mathematics teachers and also emphasised that how we teach is important, not only what we teach (Vyšín, 1980).

The interest of CZ and SK mathematicians in education and the responsibility they felt for it showed itself during the New Math movement, which appeared in the Czechoslovak context as modernisation of teaching mathematics. The Union of Czechoslovak Mathematicians and Physicists (or in some times of its history, Czech)[9] took the initiative and organised seminars and conferences (some of which

---

[8] Some parts of G. Brousseau's work (e.g., 1997) were translated into Slovak by I. Trenčanský et al. in 2011 and into Czech by J. Novotná and colleagues in 2012.

[9] This was founded in 1862 as the Verein fur freie Vorträge aus der Mathematik und Physik (Club for Free Lectures in Mathematics and Physics), and in 1869 it was renamed Jednota českých mathematiků (Union of Czech Mathematicians). It has always united both mathematicians and mathematics teachers. It was also a founding member of the European Mathematical Society. Among other activities, the Union publishes scientific journals. It started to publish *Časopis pro pěstování matematiky a fyziky* (Journal for Fostering Mathematics and Physics) in 1872 (it exists to this day under the name *Mathematica Bohemica*). In 1922, it established a journal for secondary pupils called *Rozhledy matematicko-fyzikální* (Mathematics-Physics Horizons) and in 1948 a journal for

still exist today), both in CZ and SK, where these issues were discussed. Renowned mathematicians such as E. Čech, V. Kořínek, J. Kurzweil, T. Šalát, M. Kolibiar and M. Švec supported the movement. More importantly, the Union established the Department for Modernisation of Teaching Mathematics and Physics, whose mathematical part became part of the Mathematical Institute of the Czechoslovak Academy of Sciences in 1969. The goal of the Department was to support cooperation among researchers. Its research paradigm gradually developed, which was reflected in the change of its name to the Department for the Didactics of Mathematics in the early 1980s.

In the 1960s and 1970s, many articles and books were published about modernisation that concerned not only the methods of teaching but also mathematics education. Moreover, research work in mathematics education was included in the State Plan of Fundamental Research, and, in 1965, scientific education in the theory of teaching mathematics[10] was established, which can be understood as the official beginning of scientific research in mathematics education in Czechoslovakia. Dissertation theses in this new field were to include the following sections, which, in fact, correspond to current requirements: the current state of the problem, goals, methods, results and conclusions, including recommendations for future research. Some examples of research in mathematics education before 1989 are given in Sect. 7.2.3.

After the Velvet Revolution in 1989, researchers in mathematics education from Czechoslovakia became more active internationally. The foundation of the European Society for Research in Mathematics Education (ERME) in 1997 already had Czech and Slovak participation. Some researchers became members of editorial boards of journals (such as *Educational Studies in Mathematics, Journal of Mathematics Teacher Education* and *The Mediterranean Journal for Research in Mathematics Education*) and started to work on common projects with researchers from abroad; important international conferences were organised in CZ (SEMT, 1991–2017; ERCME, 1997; PME, 2006; ESU5, 2007; CIEAEM, 2006; CERME, 2001 and 2015, and YERME Summer School, 2004 and 2016). Researchers became members of international teams working on a common topic such as the Learner's Perspective Study (see Sect. 7.3.4) or the Lexicon Project (Clarke et al., 2016).

To sum up, in this section, we saw one of the four features identified as relevant for European didactic traditions, namely the importance of the strong connection of mathematics education research with mathematics and mathematicians (see Sect. 1.2). Czech and Slovak mathematicians indeed felt responsibility for the teaching of mathematics, were authors of textbooks, participated in the education of mathematics teachers and actively helped to introduce New Math principles in the former Czechoslovakia. Moreover, one of the roots of mathematics education research is based in the Institute of Mathematics of the Czech (Czechoslovak) Academy of Sciences, with which the first researchers in mathematics education were affiliated.

---

mathematics teachers, *Matematika a fysika ve škole* (Mathematics and Physics in School), which still exists under a slightly different name. Articles from these journals are currently freely available online in a digital mathematical library (www.dml.cz).

[10]Until then, researchers had to get academic degrees in mathematics or pedagogy.

7.2.3   Czechoslovak Research in Mathematics Education
Before 1989

An important feature of emerging research in mathematics education was, probably thanks to its growth from practical issues (to verify new ways of teaching within the modernisation movement), the idea that research on teaching and curricular research had to be carried out in parallel. Changes in teaching practice needed to be combined with proper and long-term research (e.g., Kraemer, 1986). It was felt that research must be done both on the theoretical and practical levels:

> 'fundamental research in teaching mathematics' and 'the didactics of mathematics' can be seen as two fields living in a tight symbiosis.... Fundamental research is, in fact, *experimental didactics* and *the theory of teaching mathematics* is enriched by its results. On the contrary, when doing fundamental research, all present results from the theory and practices of teaching mathematics are used. (Vyšín, 1976, p. 582; authors' emphasis)

Thus, research focused on the construction, implementation and evaluation of curricula. Experimental textbooks were written and implemented at least three times and their use was rigorously evaluated, in the spirit of the present idea of design experiments.

For example, at the time of modernisation, new teaching texts were written for Grades 1–3 and used in teaching at experimental schools established throughout Czechoslovakia. Teachers were not only educated to master the new material but also to acquire new teaching methods. Pupils were given tests to see what they learned from the new texts and results were mostly elaborated in a quantitative way (with some exceptions involving qualitative research through interviews with pupils). A 100-page research report was written by researchers from the Department of the Didactics of Mathematics of the Mathematical Institute of the Czechoslovak Academy of Sciences (such as J. Kittler, M. Koman, F. Kuřina and M. Tichá), depicting the theoretical background and course of the teaching experiments with their results. A substantial part of the report consisted of conclusions highlighting necessary changes in the teaching material and its implementation and recommendations for further research. This report was reviewed by four reviewers and openly defended before stakeholders (including teachers) in 1973. Similarly, a report which concerned Grades 6–9 was defended in 1977 and a report for Grades 1–5 in 1987. Alongside this more substantial research, small-scale experiments were carried out that aimed at specific topics such as geometry, number sense and assessment (e.g., Vyšín, 1972).

It must be stressed, however, that there were two branches of research before 1989. The first was applied research realised by the Research Institute of Pedagogy and by the Research Institute of Vocational Education: Its focus was on the change of the curriculum. Fundamental research was realised by the Czechoslovak Academy of Sciences (including the Mathematical Institute with its Department of the Didactics of Mathematics[11]) and by some universities, and attempts were made to connect it

---

[11] It had a small number of its own researchers but cooperated with a number of researchers from universities and with mathematics teachers.

to applied research. The above-mentioned duality probably caused these efforts not to be entirely successful. There was no adequate reaction to the results of New Math research abroad and of local fundamental research that pointed out weak points of the New Math movement. Teaching based on set theory that influenced the West in the 1960s was introduced to schools in Czechoslovakia in 1976. However, at the same time, in schools connected to the Department of the Didactics of Mathematics, new teaching texts were already being prepared based on results from experimental teaching between 1965 and 1972 (as given in the reports mentioned above), and teaching with a set as a central concept was gradually abandoned there.

In the next period of fundamental research, when a new model of mathematical education was being sought, more attention was paid to psychological and peda-gogical aspects. Its idea was that the experience pupils had before coming to school should be used more and the style of work at school should be more active. Pupils should be encouraged to work with both non-mathematical and mathematical mod-els to acquire deep understanding of mathematical concepts. In school mathematics, there was a shift from syntactic (structural) to semantic (genetic) conceptions. For example, the ideas of assigning, dependence and variability became central when teaching functions. Explicit connections were made between school mathematics, the real world and other school subjects (Koman & Tichá, 1986, 1988).

A two-way connection between mathematics and real-world issues was in the heart of 'task environments' elaborated by J. Vyšín and M. Koman. Their focus was not on the application of mathematics in pupils' worlds only but also on impulses coming from the pupils' worlds for the building of mathematics. Vyšín (1973) emphasised that we should not teach the application of mathematics but rather mathematics that can be applied. Interestingly, Freudenthal (1991) says something similar: 'Apply-ing mathematics is not learned through teaching applications. The so-called applied mathematics lacks mathematics' greatest virtue, its flexibility. Ready-made applica-tions are anti-didactical.' (p. 85)

At the beginning of the 1980s, the main areas and characteristics of experimental teaching and the conception of fundamental research for the period 1981–1990 were formulated by the members of the Department of the Didactics of Mathematics. These consisted, among others, of lowering factual teaching, elaborating a psychological-genetic approach, more focus on mathematical methods (problem-solving methods), so-called mathematical laboratories[12] and problem teaching, and conducting research on introducing calculators in teaching mathematics. On the state level, teams of 5–10 researchers from universities, the Mathematical Institute and schools were formed, and each team suggested a research problem the team wanted to solve. Some examples were a team led by M. Koman that studied functions (e.g., Koman & Tichá, 1986), a team led by F. Kuřina that focused on geometry (e.g., Kuřina, 1976;

---

[12]Mathematical laboratories consisted of methods that helped pupils become active and in which a teacher encouraged their discovery, experimenting etc. Nowadays, we would speak about inquiry teaching as a constructivist approach to teaching. Mathematical laboratories remained mostly in theory and existed in experimental schools affiliated with the Mathematical Institute.

Koman et al., 1986) and a team led by M. Hejný that developed the methodology of research (Hejný et al., 1988).

Before 1989, important sources of new ideas in mathematics education were dissertation theses in the above-mentioned theory of the teaching of mathematics. An example is M. Tichá's unpublished dissertation from 1982 called *To Strategies of Problem Solving in Teaching Pupils Mathematics at the Lower Secondary School*. It concerned the evaluation of the use of a pilot text of J. Vyšín (*Propositional Forms*): Its goal was to find out whether, by a suitable organisation of learning conditions, pupils were not only able to master problem-solving methods but were also able to work creatively with these methods. The author used a mixed methodology of research that combined a written test with interviews of 65 randomly selected pupils. The work showed, among other things, that the graphic method helps pupils to solve word problems on movement and that even though the teacher usually emphasises a calculation method for these problems (a system of equations), pupils use insight (common sense) first, experiments next and finally equations. Research on pupils and their reasoning started to be a major focus.

A work unique in its size, character and impact was published by Slovak researchers, M. Hejný and colleagues in 1988.[13] It is a comprehensive book addressed to teachers, teacher educators and researchers. It covers all parts of secondary school mathematics in a mathematics-didactic way. Not only are there suggestions for teaching the appropriate subject matter, these suggestions are also documented by teaching experiments and interviews with pupils and teachers and augmented by mathematical problems to solve (some of which are quite complex). The didactic elaboration of topics is framed by a concept-development theory called later the theory of generic models (see Sect. 7.3.2). The authors managed to connect the mathematical, methodical, pedagogical and psychological aspects of mathematics education in a way that has inspired new researchers since. The book is still used for the education of future mathematics teachers and Ph.D. students in mathematics education.

Until 1989, the tendency in CZ and SK research went through a series of changes of focus: a one-sided emphasis on mathematical content, the contribution of mathematical education to the education of pupils and teachers, the teachers' and pupils' role in the teaching-learning process, pupils' mathematical culture, the professionalization of teachers' work, and similar aspects. After the Revolution, the character of research work has changed in the same way as opportunities have arisen such as access to international literature and research communities or grant projects. However, the experience from fundamental research realised before 1989 positively influenced mathematics educators' competence to carry out research. Unlike in most fields of didactics in CZ and SK, in which research had to be established from scratch after 1989, research in mathematics education has never really been disrupted, regardless of the difficulties researchers met under the communist regime.

---

[13] As the leading author had problems under the communist regime and could not publish freely, the book was published as a second volume to show that it was a sequel to the existing book *Teória vyučovania matematiky 1* and thus must be published. But, in fact, it is a stand-alone book, not connected to its namesake.

To sum up, in this section we saw two more of the four features identified as relevant for European didactic traditions (see Sects. 1.4 and 1.5). First, a key role of design activities for learning and of teaching environments can already be seen in the design and testing of materials for the New Math movement, and it remains so until today. In our tradition, we can also distinguish an empirical turn similar to the German one (see Sect. 5.2.3), from '*Stoffdidaktik*' focusing strongly on mathematical content for direct use in lessons to design activities done to study the effect of didactic variables in classroom experiments. Again, the New Math movement greatly contributed to this change. Second, another feature of the Czechoslovak tradition is the basis in empirical research using various research methods. The research is usually on a small scale, and before 1989 it was strongly connected to the experimental primary schools mentioned above.

7.3 Mathematics Education Research in the Czech Republic and in Slovakia After 1989

Since the political change in 1989, the field of mathematics education has developed against a background of big changes in education and teacher education brought about by reform efforts. These were mostly promoted by educationalists, so naturally, many of Comenius's ideas can be found in their background in CZ. For example, the key policy document, the *National Programme for the Development of Education in the Czech Republic*, the so-called *White Paper* (2001), strives to revive Comenius's plea for universal education and brings forward principles that would ensure an access to education to every single individual regardless of age, class, gender and nationality. Even more profound was the (unvoiced) influence of Comenius's ideas on the main curricular documents, the Framework Educational Programmes, mainly in the conception of key competences that the school should develop. For example, learning competencies and problem-solving competencies are based on Comenius's principle of systematicity, autonomy and activity (Smílková & Balvín, 2016). The other competencies (communication, social and personal, civil and working competencies) are directly linked to Comenius's principles. The same is true for cross-curricular subjects that are introduced in the documents and that make an inseparable part of basic education and represent its important formative element, namely, personal and social education, democratic citizenship, education towards thinking in European and global contexts, multicultural education and environmental education.

Comenius and his work are alive in CZ, not only in works of researchers from the Department of Comenius Studies in the Czech Academy of Sciences. His ideas are not only embedded in textbooks for university studies of future teachers but also in disciplines such as social pedagogy, philosophy and religion. Comenius's influence in mathematics education can be traced as well even though it is not always stated; his ideas are embodied in the milieu within which researchers work. Researchers often refer to Comenius's work when promoting visualisation in teaching, learning by doing and by using several senses, learning by playing, and the like, or life-long learning.

7.3.1 Methodology

In the last 25 years or so, research in former Czechoslovakia has been rather diversified. The common history of the Czech Republic and Slovakia was disconnected in 1993 when Czechoslovakia was separated into two countries: the Czech Republic and the Slovak Republic, who have had independent developments. Due to the same roots, educational systems and teacher education in both countries are still close, and naturally there is cooperation among researchers from both countries. Thus, we can present research in both countries together.

Roughly, the research can be divided into three branches. The first one is orientated towards the study of thinking processes of pupils and teachers, communication in teaching and learning mathematics, climate of the classroom, and the whole socio-cultural context. The second branch is on investigating curriculum, mathematical content, textbooks and so on. Obviously, these two branches are closely connected, even though sometimes the connection is not considered in CZ and SK research. The third branch focuses on the history of mathematical ideas and strives to find inspirations for mathematical education in it. This chapter mostly concentrates on the first branch because its basis is mainly empirical research, and quality publications in English are available. It will be divided into four streams, in which main research studies will be briefly summarised and illustrated by representative examples of publications. We have given priority to English or German publications originating from a particular research study (if they exist at all) over Czech or Slovak ones even when the latter are newer or more comprehensive so that an international audience can have access to them. Naturally, some research sits at the border of our identified streams; for example, sometimes pupils' reasoning is studied in relation to teachers' knowledge. Nevertheless, we decided to make these distinctions for the sake of clarity.

When trying to distinguish main research topics and to find relevant publications by CZ and SK authors, we scrutinised scientific journals, proceedings of international conferences and books to which we had access. We searched the database of prestigious research projects awarded by the Grant Agency of the Czech Republic (GA ČR) and its Slovak counterpart, the Scientific Grant Agency of the Ministry of Education of the Slovak Republic (VEGA). For the Czech part, we also used results of a survey conducted in 2013 at all universities educating future mathematics teachers done by the former Accreditation Committee established by the Ministry of Education, Youth and Sport. In this survey, the universities were asked to include the most important publications on research in mathematics education written by their employees. For the Slovak research, we contacted our research colleagues from Slovakia to direct us to results of Slovak research that we were not familiar with.

We do not present work done by Czech or Slovak researchers who have gone to work abroad (such as J. Višňovská or J. Trgalová).

We realise that our description of the state of affairs must necessarily be incomplete. Research publications may result outside of awarded grant projects or outside of universities educating mathematics teachers. Moreover, we base our considerations on published results only and mostly on empirical research. There may be on-going research without any publication known to us. Finally, we would like to stress that the research described below should be understood as a continuation of research up to the Velvet Revolution as described in previous sections.

7.3.2   Development of Theories

Another of the identified main features of European traditions (see Sect. 1.3) is the key role of theory. Unlike the French tradition, in which research in mathematics education has evolved around three basic theories (see Sect. 2.1.2), the situation in CZ and SK is more diversified. However, we can say that most research is based on the constructivist theory of learning, which has been elaborated in the local context as didactic constructivism (Hejný & Kuřina, 2009). M. Hejný and F. Kuřina formulated 10 principles of didactic constructivism that have influenced teaching, teacher education and research in mathematics education in CZ and to a lesser extent in SK. One of the principles is an emphasis on pupils' mathematical activity consisting of looking for things such as relationships, problem solving, generalising and argumentation. Another is the creation of the kind of environment in a lesson that supports creativity and is the basis of learning mathematics. The principles also stress the importance of pupils' mistakes and the way teachers can handle them to develop pupils' knowledge. This aspect has affected both research and teaching in CZ and SK.

In terms of concept-development theories, by far the most influential one in CZ and SK is the theory of generic models that was originally developed by M. Hejný in Bratislava in the 1980s and later in Prague. Unfortunately, the theory is not described in its entirety in English. The most comprehensive book about it is in Czech (Hejný, 2014). Elements of the theory are described in various publications such as Hejný (2012), Hejný and Kuřina (2009) and Stehlíková (2004). In brief, the theory describes concept development in mathematics as consisting of several levels, beginning with motivation, through the stage of isolated models (concrete cases of future knowledge) and the stage of generic models (which comprise all isolated models and can substitute for them) up to the abstract knowledge level.[14] There are two shifts between the stages: generalisation and abstraction. The latter is accompanied by a change in

---

[14]For example, when pupils solve a problem on generalising a pattern represented by several numbers, they first calculate several other elements of the sequence and thus work with isolated models. Later, they can see a rule and are able to use it for the calculation of further elements: They have found a processual generic model. Next, they are able to state in words how any element of the sequence can be found without having to calculate the preceding elements: They have found a

language (for example, the language of algebra is used). Crystallisation is the term used for the process of connecting new concepts to old ones and using it to build new knowledge in the future. Within this theory, insufficient understanding is captured by the term *mechanical understanding*, which means knowledge that is not supported by generic models and is mostly grasped by memory only.

This theory has been successfully used by researchers in CZ, SK and Poland for the description of the construction of knowledge from different fields of mathematics for pupils and students of different ages (e.g., Jirotková & Littler, 2002; Jirotková & Slezáková, 2013; Krpec, 2016; Robová, 2012; Stehlíková, 2004; Vaníček, 2009).[15] On the one hand, the theory has practical applications, the most prominent being a new approach to teaching called scheme-based education (Hejný, 2012). On the other hand, L. Kvasz embraced the theoretical foundations of the theory and the teaching style based on it by formulating the principles of so-called genetic constructivism (Kvasz, 2016) to show that it differs from radical constructivism with which it is sometimes identified in the local context. L. Kvasz grounds his arguments in the genetic approach to mathematics, which is based on a thorough understanding of the history and epistemology of mathematics.

While the theory of generic models had to be described here, as it is mostly rooted in CZ and SK, the other theories which are often used by Czech and Slovak researchers do not require such a description. Very influential is the theory of didactical situations developed by Brousseau (1997). The theory is mostly used by J. Novotná and her collaborators in CZ and by I. Trenčanský, L. Rumanová and E. Smiešková in SK. For the research aimed at future teachers, Shulman's (1986) theory is mostly used as well as Ball and colleagues' practice-based theory of mathematical knowledge for teaching (Hill et al., 2008). Research focusing on a teacher's use of technology is carried out against the background of technological pedagogical content knowledge (TPACK; Mishra & Koehler, 2006).

7.3.3   Knowledge and Education of Future Elementary and Mathematics Teachers

Knowledge and education of future elementary and mathematics teachers have attracted much attention in CZ and SK; however, much of it has been in the form of theoretical studies or recommendations, which are not our focus here. In terms of empirical research, one strand of research aims at mathematical knowledge for teaching of future elementary teachers (Marcinek & Partová, 2011; Partová et al., 2013; Samková & Hošpesová, 2015), for example, knowledge about geometric shapes, which has been investigated within a VEGA project aimed at geometric conceptions

---

conceptual generic model. When they are able to write an algebraic expression for the nth term of the sequence, they are at the abstract level.

[15] The theory has also been used in many dissertation and diploma theses.

and misconceptions of both pre-school and school age children and future teachers (Duatepe-Paksu et al., 2017; Žilková et al., 2015).

Another strand aims at pedagogical content knowledge of future teachers. First, problem-posing competence of future teachers, understood as an educational, motivational and diagnostic tool, has been studied in Tichá and Hošpesová (2013). Their analysis of the problems posed by the students revealed, among other things, shortcomings in their conceptual understanding of some notions, especially fractions. Classroom-based joint reflection became the means of re-education. Problem posing of pre- and in-service teachers and the way they reflect on the posed problems has been a focus of Hošpesová and Tichá (2015). They have confirmed, among other things, that problem posing on its own is by no means a sufficient tool for the remedy of teachers' misconceptions. It works best in combination with reactions from others. Some ways have been shown in which teacher educators can guide joint reflections to achieve best results.

The second focus is on a professional vision of future teachers, both in terms of their patterns of attention in general and attention to mathematics-specific phenomena in particular (Vondrová & Žalská, 2015), and possibilities for the development of a professional vision in a control versus experimental group intervention study (Simpson, Vondrová & Žalská, 2018). While the latter study confirmed many results of research on professional vision of future mathematics teachers in terms of their pattern of attention (for example, more attention to the teacher than to the pupils, more attention to pedagogical issues rather than to issues connected with the teaching of mathematical content), it showed a markedly different development in student teachers' knowledge-based reasoning (unlike in related literature, there was no shift towards interpretation of noticed events and issues).

The third focus is on future teachers' TPACK or its aspects (Beňačka & Čeretková, 2015; Jančařík & Novotná, 2013; Kapounová et al., 2013). Finally, we will mention skills and knowledge that future teachers should possess when teaching content and language-integrated learning (CLIL), which refers to any learning context in which content and language are integrated in order to fulfil specified educational aims (Marsh & Langé, 1999). Moraová and Novotná's (2005, 2017) research is an example of examining ways of introducing this teaching strategy to future teachers.

Teachers' teaching practices, beliefs and ideas are also topics of research. For example, their views of what they consider critical parts of mathematics for their pupils' understanding and how they deal with them in their teaching were the centre of attention in a GA ČR project (Rendl et al., 2013). According to teachers, pupils tend to make more mistakes when more rules and procedures are learnt. When the knowledge is needed for a task at a later time, there can be interference among the variety of knowledge already learnt, and pupils have more tendency to fail if the task does not include an explicit reference to the necessary knowledge and instead the knowledge is only implied. The solution teachers have offered consists of further revision and drill; however, it is not clear whether an intended goal of the revision and drill is also to deepen conceptual understanding or whether there is an overreliance on its spontaneous emergence.

Teachers' educational styles are investigated in Hejný (2012) and Jirotková (2012). The authors have developed (and applied) a diagnostic tool that can be used for the characterisation of a teacher's educational style. The tool consists of 20 parameters divided into four areas: beliefs, experience, personality and abilities/competences. The tool can be used for the development of the teacher's teaching style as well.

Collective reflection has been studied as a means of influencing teachers' beliefs. The springboard of research on reflection was the cooperation with elementary teachers that naturally led to action research and the study of its various conceptions, including its importance for collaborative continuing professional development (Benke et al., 2008). Reflection is understood not only as one of the competences but also as one of the ways of developing this competence. Tichá and Hošpesová (2006) based their research on samples of teachers' reflections and showed their gradual development from merely being simple conversations based on intuitive perceptions through searching for effective teaching approaches, to the deep assessment of mathematics teaching from the point of view of topics and their didactic elaboration, and to suggestions for the teachers' own experiments.

The cooperation between researchers and teachers is mainly focused on case studies (Jirotková, 2012; Tichá & Hošpesová, 2006), which document the developments of teachers' knowledge and beliefs. The question of teachers as researchers is investigated, for example, in Novotná et al. (2003) and in a broader context in Novotná et al. (2013). The latter work focuses on teacher education and its development and introduces, among other things, several categories of mathematics educators. Observations are discussed as important means of learning in both teachers' and researchers' practices.

7.3.4   Classroom Research

An interest in classroom research naturally stems from the need to understand situations that appear in the classroom and that enable the development of pupils' mathematical knowledge.

First, we mention extensive classroom research within the Learner's Perspective Study project, whose goal has been to conduct international comparative studies of teaching mathematics. Its characteristic feature is that 10 successive lessons are video-recoded, described by artefacts and complemented by interviews with the teacher and with some pupils. Results are published in a series of books in which Czech researchers also have chapters (Binterová et al., 2006; Novotná & Hošpesová, 2010, 2013, 2014). The chapters include analyses of teaching situations from both national and international perspectives. They bring insight into the outer and inner stereotypes in lower-secondary mathematics lessons from the viewpoint of, for example, classroom environment, algebraic competence and its development, coherence between educational goals and school practice when teaching word problems, active involvement of pupils in concept development and the like.

Much research of processes in mathematical lessons has been conducted in CZ in connection with doing trials of a specific way of teaching mathematics at the primary school level mentioned above, the so-called scheme-based education[16] which is developed in the research team around M. Hejný (Hejný, 2012, 2014; Jirotková & Slezáková, 2013). This teaching is based on the above theory of generic models. Mathematical knowledge is built based on pupils' active work within carefully developed learning environments that permeate all grades of the primary school. They meet the requirements of substantial learning environments as introduced by Wittmann (1995). The teacher's role in teaching mathematics is that of a facilitator. The studies of scheme-based education are of an exploratory nature, bringing insight into how pupils reason in mathematics during a lesson conducted in the frame of scheme-based education (e.g., Jirotková & Slezáková, 2013; Krpec, 2016).

Classroom research also plays an important role in three research projects awarded funding by GA ČR. The first investigated the development of mathematical literacy at a primary school (Hošpesová et al., 2011). Among other things, the project provided reasons why we should emphasize links between and blending of a teacher's mathematical competences and didactic issues in that it justifies an emphasis on systematic practice in which pupils construct their own mathematical world. The project demonstrated the importance of the role that arithmetical, algebraic and geometric models play in the development of notions of mathematical objects and relations between them and in solving problems stemming from real-life situations.

The second project focused on the use of textbooks (not only in mathematics) at lower-secondary schools (Sikorová, 2011). It showed, for example, that practising mathematics teachers used the textbook mostly as 'a source of the tasks' and for simple activities such as reading the text.

The third project aimed at the development of the culture of solving mathematical problems in school practice. Its focus was pupils' solving strategies, their reasoning, mistakes and the use of research results in school practice (Eisenmann et al., 2015). The research was longitudinal. Lower- and upper-secondary pupils were exposed to the use of selected heuristic strategies in mathematical problem solving for a period of 16 months. A tool was developed that allows for the description of their ability to solve problems and that consists of four components: intelligence, text comprehension, creativity and the ability to use existing knowledge. After the experimental teaching, the pupils appeared to considerably improve in the creativity component and there was a positive change in their attitude to problem solving. The teaching style of the teachers participating in the experiment showed a significant change towards a more constructivist, inquiry-based approach and in their willingness to accept a pupil's non-standard approach to solving a problem.

Finally, we will mention intervention studies of an experimental versus control group type, which are not frequent in published CZ and SK research[17] (e.g., Sedláček,

---

[16]A scheme is understood in the sense of Gerrig (1991) as a memory structure that incorporates clusters of information relevant to comprehension. Hejný's conception of a scheme builds on Piaget's conception and is close to Dubinský and McDonald's (1999).

[17]Many studies only include an experimental group (e.g., Bero, 1993; Binterová & Fuchs, 2014).

2009; Slavíčková, 2007). In Cachová's (2011) study, the experimental group of Grade 1 pupils working with calculators performed the same as the control group working without them in tasks and were more motivated than the control group. In Huclová and Lombart's study (2011), the experimental group of pupils working with software only when learning about line symmetry picked up some unsuitable construction steps based on the way the software made them perform these steps. Vankúš (2008) investigated game-based learning mathematics and found that it improved pupils' motivation and attitudes to mathematics but that there was no difference between the experimental and control groups in terms of mathematics knowledge acquired.

7.3.5 Pupils' Reasoning in Mathematics

Probably most attention in CZ and SK research has been devoted to pupils' strategies, reasoning, mistakes and misconceptions in concrete parts of mathematics. Some examples of these areas of study are fractions (Tichá, 2000), geometric reasoning and conceptions (Budínová, 2017; Kopáčová & Žilková, 2015; Pavlovičová & Barcíková, 2013), non-sighted pupils' perception of space and its objects (Kohanová, 2007), word problems (Eisenmann et al., 2015; Hejný, 2006; Tichá & Koman, 1998), solids (Jirotková & Littler, 2002), infinity (Cihlář et al., 2015; Jirotková, 1998), measurement in geometry (Tůmová & Vondrová, 2017), representations of multiplication (Partová & Marcinek, 2015) and combinatorial problems strategies (Janáčková, 2006).

A GA ČR project (see Vondrová et al., 2015) focused on interviews with pupils solving problems from critical areas of mathematics as identified by teachers (see Sect. 7.3.3). Research revealed, among other things, a strong tendency of pupils towards the use of formulas in geometry and preference to calculations over reasoning (which was not mentioned by the interviewed teachers). Deficiencies were identified in the mental representation of a continuum of rational numbers, the conceptual understanding of an algebraic expression as an object to be manipulated (as opposed to procedural problems reported by the teachers), the breach of relationships between theoretical and spatial-graphic spaces when interpreting and using a picture in geometry, and the conceptual understanding of measure in geometry. A serious problem was identified in word problems. There was no direct correspondence between some language expressions and their mathematical descriptions in numbers and variables. A relationship between language literacy and problem solving is currently being investigated within a new GA ČR project aimed at parameters influencing the difficulty of word problems where only preliminary results have been published (e.g., Vondrová, Novotná & Havlíčková, 2018).

On the one hand, research attention has been focused on pre-school children and their reasoning. For example, problem solving and the diversity of pupils' solutions (Kaslová, 2017), geometric ideas (Kuřina, Tichá & Hošpesová, 1998), and children discovering mathematical concepts and strategies in a learning environment (Jirotková & Slezáková, 2013). On the other hand, the development of mathematical knowledge of university students has also been studied (Simpson & Stehlíková, 2006; Stehlíková, 2004). For example, the former study documented that in the development of an examples-to-generality pedagogy, an emphasis on the guidance of joint attention is needed rather than the free-for-all of unguided discovery, that is, on teachers and learners making sense of structures together, with the teacher able to explicitly guide attention to those aspects of the structure that will be the basis of later abstraction and to the links between the formal and general with specific examples.

A specific place in this stream of research is held by work on the use of technology in mathematics education. Many studies on this topic written by CZ and SK authors have focused on a mathematical content and how it can be presented "better" or "more easily" via technology. Mathematical problems have been suggested that can be used in, for example, GeoGebra and sometimes given a trial with a small number of participants. Such studies may function as springboards for empirical research and are valuable in their own right for actual teaching. However, as they are not research reports as such, we will not address them further here.

Two Czech comprehensive books on technology have influenced the field in CZ. Vaníček (2009) presents ways dynamic geometry software can enhance the teaching of mathematics and points to some perils of the software, using personal results of teaching experiments and experience from teacher education. Against the background of pupils' active learning of mathematics, Robová (2012) analysed results of research on the use of ICT and presented a comprehensive account of the merits and perils of ICT use in the teaching of mathematics. For the first time in CZ research, she focuses on the use of internet material and forums in teaching. She argues that the successful use of ICT must be accompanied by a change in a teacher's teaching styles, which is often not the case. Moreover, in Robová (2013), she presents skills that pupils should develop to overcome the problems she identified in their use of ICT tools. Examples are estimation skills, understanding dependence and congruence of objects in a dynamic software, and zooming in and out.

7.4 Current Situation, Perspectives and Challenges

In this chapter, we strove to provide an international reader with a glimpse of the current trends in research in mathematics education in the Czech Republic and in Slovakia and the roots from which they stem. One of these can be found within Comenius's principles. Comenius's influence can be traced not only in mathematics education but in education in general. However, many of his ideas are taken for granted by researchers in CZ and SK and often used without explicit reference to their proponent. They are also well known among practising teachers, though more often than not on the level of proclamation.

In CZ, research in mathematics education has mostly been concentrated around faculties preparing teachers and in the Institute of Mathematics of the Czech Academy of Sciences. Nowadays, mathematics education is recognised as a scientific field. There are doctoral studies in mathematics education and researchers can reach habilitation or professorship in this field. Researchers have ample opportunities to present their work at international conferences. There has been an effort to bring together researchers from the didactics of different subjects to get new insights and develop a common scientific language, with a view that fields with developed didactics such as mathematics help the fields that have been newly established. One of the indicators of this effort was the establishment of a scientific journal aimed at mathematics, physics and sciences education: *Scientia in educatione* (www.scied.cz).

In addition, research in mathematics education must face old and new challenges. The challenges include an insufficient number of grant projects to be awarded to researchers in education (as opposed to researchers in sciences or medicine), a growing pressure on researchers to publish all the time (so that they put a lot of energy into publishing partial results; prefer small-scale, short-term studies and do not have enough time to devote to publishing studies aimed at using research results in practice), insufficient financial means for doctoral students, and a lack of career incentives for them, which has had effects such as a decrease in the number of Ph.D. students, and research in mathematics education not developing equally in all universities educating teachers.

In terms of mathematics education research in CZ and SK, we can see, for example, the need to conduct long-term empirical research investigating the influence of certain types of teaching (such as the above scheme-based teaching of mathematics, which has been used by an increasing number of schools in CZ) on pupils' mathematical knowledge. More attention is needed on research on the upper-secondary school level, as research of elementary and lower-secondary school levels prevails.

Much work in mathematics education in CZ and SK has investigated mathematical content and its elaboration for teaching (especially in the context of teaching with the help of technology) without taking into account pupils, teachers and the whole socio-cultural context of the class and society. Proper methodology is needed. Moreover, it seems to us that some studies do not build on existing research results in the investigated topic, without which the field will not advance.

Another possibility for development can be seen in mutual cooperation among researchers from different disciplines. While there has been enough contact with mathematics, cooperation with pedagogy and psychology is needed. Teams of researchers from these fields working on a common topic are not frequent in CZ and SK.[18]

Finally, let us reiterate that even though we have tried to be as rigorous as possible and ground the above account in publications available to us, the text naturally represents our own personal perspective.

Acknowledgements The article was supported by research Progress Q17 Teacher Preparation and the Teaching Profession in the Context of Science and Research (Novotná and Vondrová) and by RVO: 67985840 (Tichá).

---

[18]There have been some exceptions, for example, researchers from mathematics education, psychology and linguistics cooperating within a GA ČR project on word problems, researchers from linguistics and mathematics education working on CLIL issues and researchers from different fields investigating professional vision of student teachers and possibilities for their development.

References

Benke, G., Hošpesová, A., & Tichá, M. (2008). The use of action research in teacher education. In K. Krainer & T. Wood (Eds.), *Participants in mathematics teacher education; individuals, teams, communities and networks (The international handbook of mathematics teacher education* (Vol. 3, pp. 283–330). Rotterdam: Sense Publishers.

Beňačka, J., & Čeretková, S. (2015). Graphing functions and solving equations, inequalities and linear systems with pre-service teachers in excel. In K. Krainer & N. Vondrová (Eds.), *Proceedings of the Ninth Congress of the European Society for Research in Mathematics Education* (pp. 2311–2318). Prague, Czech Republic: Charles University in Prague, Faculty of Education and ERME.

Bero, P. (1993). Calculations in the style of Kepler. *For the Learning of Mathematics, 13*(3), 27–30.

Binterová, H., & Fuchs, E. (2014). How to teach mono-unary algebras and functional graphs with the use of computers in secondary schools. *International Journal of Mathematical Education in Science and Technology, 45*(5), 742–754. https://doi.org/10.1080/0020739x.2013.877604.

Binterová, H., Hošpesová, A., & Novotná, J. (2006). Constitution of classroom environment: Case Study. In D. Clarke, Ch. Keitel & Y. Shimizu (Eds.), *Mathematics classrooms in twelve countries: The insider's perspective* (pp. 275–288). Rotterdam: Sense Publishers.

Brousseau, G. (1997). *Theory of didactical situations in mathematics 1970–1990* (N. Balacheff, M. Cooper, R. Sutherland & V. Warfield, Eds. & Trans.). Dordrecht: Kluwer Academic Publishers.

Brousseau, G. (2011). *Základy a metódy Didaktiky matematiky [Basics and methods of mathematics education]*. (I. Trenčanský et al., Eds. & Trans.). Bratislava: Univerzita Komenského, FMFI.

Brousseau, G. (2012). *Úvod do Teorie didaktických situací v matematice [Introduction into the Theory of Didactical Situations in Mathematics]*. (G. Brousseau, J. Novotná, J. Bureš, L. Růžičková, Trans.). Prague: Univerzita Karlova v Praze – Pedagogická fakulta.

Budínová, I. (2017). Progressive development of perception of the concept of a square by elementary school pupils. In J. Novotná & H. Moraová (Eds.), *Proceedings of SEMT '17. Equity and Diversity in Elementary Mathematics Education* (pp. 109–118). Prague: Charles University, Faculty of Education.

Cachová, J. (2011). Kalkulačka v elementární aritmetice [Calculator in elementary arithmetic]. In A. Hošpesová, F. Kuřina, J. Cachová, J. Macháčková, F. Roubíček, M. Tichá & J. Vaníček (Eds.), *Matematická gramotnost a vyučování matematice* (pp. 111–150). České Budějovice: Jihočeská univerzita.

Cihlář, J., Eisenmann, P., & Krátká, M. (2015). Omega Position – A specific phase of perceiving the notion of infinity. *Scientia in educatione, 6*(2), 51–73. http://scied.cz/index.php/scied/article/view/184. Accessed 8 January 2018.

Clarke, D., Díez-Palomar, J., Hannula, M., Chan, M. C. E., Mesiti, C., Novotná, J. ... Dobie, T. (2016). Language mediating learning: The function of language in mediating and shaping the classroom experiences of students, teachers and researchers. In C. Csíkos, A. Rausch & J. Szitányi (Eds.), *Proceedings of the 40th Conference of the International Group for the Psychology of Mathematics Education* (Vol. 1, pp. 349–374). Szegéd: PME.

Duatepe-Paksu, A., Rybanský, L´., & Žilková, K. (2017). The content knowledge about rhombus of Turkish and Slovak pre-service elementary teachers. In J. Novotná & H. Moraová (Eds.), *Proceedings of SEMT '17. Equity and Diversity in Elementary Mathematics Education* (pp. 158–168). Prague: Charles University, Faculty of Education.

Dubinsky, E., & McDonald, M. (1999). APOS: A constructivist theory of learning in undergraduate mathematics education research. In D. Holton (Ed.), *The teaching and learning of mathematics at university level: An ICMI Study* (pp. 275–282). Dordrecht: Kluwer Academic Publishers.

Eisenmann, P., Novotná, J., Přibyl, J., & Břehovský, J. (2015). The development of a culture of problem solving with secondary students through heuristic strategies. *Mathematics Education Research Journal, 27*(4), 535–562.

Freudenthal, H. (1972). *Mathematics as an educational task*. Dordrecht: D. Reidel Publishing Company.

Freudenthal, H. (1986). *Didactical phenomenology of mathematical structures*. Dordrecht: Kluwer Academic Publishers.

Freudenthal, H. (1991). *Revisiting mathematics education. The Netherlands*. Dordrecht: Kluwer Academic Publishers.

Gerrig, R. J. (1991). Text comprehension. In R. J. Sternberg & E. E. Smith (Eds.), *The psycholgy of human thought* (pp. 244–245). Cambridge: Cambridge University Press.

Hejný, M. (2012). Exploring the cognitive dimension of teaching mathematics through scheme-oriented approach to education. *Orbis scholae, 6*(2), 41–55.

Hejný, M. (2014). *Vyučování orientované na budování schémat: Aritmetika 1. stupně [Teaching oriented at building schemes: Elementary arithmetic]*. Prague: PedF UK.

Hejný, M., Bálint, L., Benešová, M., Bereková, H., Bero, P., Frantiková, L. … Vantuch, J. (1988). *Teória vyučovania matematiky 2 [The theory of teaching mathematics 2]*. Bratislava: SPN.

Hejný, M. (2006). Diversity of students' solutions of a word problem and the teachers' educational style. In A. Simpson (Ed.), *Retirement as process and concept a festschrift for Eddie Gray and David Tall* (pp. 109–117). Prague: Charles University, Faculty of Education.

Hejný, M., & Kuřina, F. (2009). *Dítě, škola, matematika. Konstruktivistické přístupy k vyučování [Child, school, mathematics. Constructivist approaches to teaching]*. Prague: Portál.

Hill, H. C., Blunk, M. L., Charalambous, C. Y., Lewis, J. M., Phelps, G. C., Sleep, L., et al. (2008). Mathematical knowledge for teaching and the mathematical quality of instruction: An exploratory study. *Cognition and Instruction, 26*(4), 430–511. https://doi.org/10.1080/07370000802177235.

Hošpesová, A., Kuřina, F., Cachová, J., Macháčková, J., Roubíček, F., Tichá, M., et al. (2011). *Matematická gramotnost a vyučování matematice [Mathematical literacy and teaching mathematics]*. České Budějovice: Jihočeská univerzita.

Hošpesová, A., & Tichá, M. (2015). Problem posing in primary school teacher training. In F. M. Singer, N. Ellerton & J. Cai (Eds.), *Mathematical problem posing: From research to effective practice* (pp. 433–447). New York: Springer Science + Business Media.

Hruša, K. (1962). *Metodika počtů pro pedagogické instituty 1, 2 [Teaching methodology of counting for pedagogical institutes]*. Prague: SPN.

Huclová, M., & Lombart, J. (2011). Rizika nahrazení rýsování na papír konstruováním pomocí ICT při školní výuce geometrie [Risks in substituting constructions on paper with constructions with the help of ICT when teaching school geometry]. In *Sborník příspěvků 5. konference Užití počítačů ve výuce matematiky* (pp. 157–174). České Budějovice: Jihočeská univerzita.

Jackson, R. (2011). The origin of Camphill and the social pedagogic impulse. *Educational Review, 63*(1), 95–104. https://doi.org/10.1080/00131911.2010.510906.

Janáčková, M. (2006). How the task context influences the task resolving. *Acta Didactica Universitatis Comenianae Mathematics, 6*, 13–24.

Jančařík, A., & Novotná, J. (2013). E-didactical shift. In Kvasnička, R. (Ed.), *10th International Conference on Efficiency and Responsibility in Education, Proceedings* (pp. 240–247). Prague: Czech University of Life Sciences.

Jirotková, D. (1998). Pojem nekonečno v geometrických představách studentů primární pedagogiky [The concept of infinity in geometric images of future elementary teachers]. *Pokroky matematiky, fyziky a astronomie, 43*(4), 326–334.

Jirotková, D. (2012). A tool for diagnosing teachers' educational styles in mathematics: development description and illustration. *Orbis scholae, 6*(2), 69–83.

Jirotková, D., & Littler, G. (2002). Investigating cognitive and communicative processes through children's handling with solids. In A. D. Cockburn & E. Nardi (Eds.), *Proceedings of the 26th Conference PME* (pp. 145–152). Norwich: University of East Anglia.

Jirotková, D., & Slezáková, J. (2013). Didactic environment bus as a tool for development of early mathematical thinking. In J. Novotná & H. Moraová (Eds.), *Proceedings of SEMT '13. Tasks and Tools in Elementary Mathematics* (pp. 147–154). Prague: Charles University, Faculty of Education.

Kapounová, J., Majdák, M., & Novosad, P. (2013). Evaluation of e-learning courses for lifelong learning. In M. Ciussi & M. Augier (Eds.), *Proceedings of the 12th European Conference on e-Learning* (pp. 173–183). France: Sophia Antipolis.

Kaslová, M. (2017). Diversity of results in research in the domain of pre-school mathematics at kindergarten. In J. Novotná & H. Moraová (Eds.), *Proceedings of SEMT '17. Equity and Diversity in Elementary Mathematics Education* (pp. 255–264). Prague: Charles University, Faculty of Education.

Kohanová, I. (2007). Assessing the attainment of analytic – Descriptive geometrical thinking with new tools. In D. Pitta-Pantazi & G. Philippou (Eds.), *Proceedings of CERME 5* (pp. 992–1011). Larnaca: University of Cyprus.

Koman, M., & Tichá, M. (1986). Základní výzkum vyučování matematice na 1. stupni základní školy [Fundamenal research in the teaching of elementary mathematics]. In *Sborník semináře Matematika a výpočetní technika ve studiu učitelství pro 1.-4. ročník ZŠ* (pp. 120–133). Prague: PedF UK.

Koman, M., & Tichá, M. (1988). *Basic research in didactics of mathematics in Czechoslovakia. Comment to the development of the functional and algorithmic approaches to problem solving.* Prague: MÚ ČSAV.

Koman, M., Kuřina, F., & Tichá, M. (1986). Some problems concerning teaching geometry to pupils aged 10 to 14 years. In R. Morris (Ed.), *Studies in mathematics education: Teaching of geometry* (Vol. 5, pp. 81–96). Paris: UNESCO.

Kopáčová, J., & Žilková, K. (2015). Developing children's language and reasoning about geometrical shapes - A case study. In J. Novotná & H. Moraová (Eds.), *Proceedings of SEMT'15. Developing Mathematical Language and Reasoning* (pp. 184–192). Prague: Charles University, Faculty of Education.

Kraemer, E. (1986). Vývoj školské matematiky a didaktiky matematiky v ČSR v období 1945-1985 [Development of school mathematics and mathematical education in Czechoslovakia between 1945 and 1985]. In I. Netuka (Ed.), *Vývoj matematiky v ČSR v období 1945–1985 a její perspektivy* (pp. 184–204). Prague: Univerzita Karlova.

Krpec, R. (2016). Isomorphism as generalization tool (in combinatorics). *Didactica Mathematicae, 38,* 107–147.

Krygowská, Z. (1977). *Zarys dydaktyki matematyki 1-3.* Warszawa: Wydawnictwa Szkolne i Pedagogiczne.

Kuřina, F. (1976). *Problémové vyučování v geometrii [Problem teaching in geometry].* Prague: SPN.

Kuřina, F., Tichá, M., & Hošpesová, A. (1998). What geometric ideas do the pre-schoolers have? *Journal of the Korea Society of Mathematical Education Series D: Research in Mathematical Education, 2*(2), 57–69.

Kvasz, L. (2016). Princípy genetického konštruktivizmu [Principles of genetic constructivism]. *Orbis Scholae, 10*(2), 15–45.

Maviglia, D. (2016). The main principles of modern pedagogy in 'Didactica Magna' of John Amos Comenius. *Creative Approaches to Research, 9*(1), 57–67.

Marcinek, T., & Partová, E. (2011). Measures of mathematical knowledge for teaching: Issues of adaptation of a U.S.-developed instrument for the use in the Slovak Republic. In J. Novotná & H. Moraová (Eds.), *Proceedings of SEMT '11. The Mathematical Knowledge Needed for Teaching in Elementary Schools* (pp. 229–236). Prague: Charles University, Faculty of Education.

Marsh, D., & Langé, G. (Eds.). (1999). *Implementing content and language integrated learning.* Jyväskylä, Finland: Continuing Education Centre, University of Jyväskylä.

Mikulčák, J. (2010). Příprava učitelů matematiky [Preparation of mathematics teachers]. In J. Mikulčák (Ed.), *Nástin dějin vyučování v matematice (a také školy) v českých zemích do roku 1918* (pp. 272–284). Prague: Matfyzpress.

Ministry of Education, Youth and Sport. (2001). *National programme for the development of education in the Czech Republic. White paper.* Prague: The Institute for Information on Education.

Mishra, P., & Koehler, M. J. (2006). Technological pedagogical content knowledge: A new framework for teacher knowledge. *Teachers College Record, 108*(6), 1017–1054.

Moraová, H., & Novotná, J. (2017). Higher order thinking skills in CLIL lesson plans of preservice teachers. In J. Novotná & H. Moraová (Eds.), *International Symposium Elementary Maths Teaching SEMT '17, Proceedings* (pp. 336–345). Prague: Charles University, Faculty of Education.

Novotná, J., & Hošpesová, A. (2010). Linking in teaching linear equations - Forms and purposes. The case of the Czech Republic. In Y. Shimizu, B. Kaur, R. Huang & D. Clarke (Eds.), *Mathematical tasks in classrooms around the world* (pp. 103–117). Rotterdam: Sense Publishers.

Novotná, J., & Hošpesová, A. (2013). Students and their teacher in a didactical situation: A case study. In B. Kaur, G. Anthony, M. Ohtani & D. Clarke (Eds.), *Student voice in mathematics classrooms around the world* (pp. 133–142). Rotterdam: Sense Publishers.

Novotná, J., & Hošpesová, A. (2014). Traditional versus investigative approaches to teaching algebra at the lower secondary level: The case of equations. In F. K. S. Leung, K. Park, D. Holton & D. Clarke (Eds.), *Algebra teaching around the world* (pp. 59–79). Rotterdam: Sense Publishers.

Novotná, J., Lebethe, A., Rosen, G., & Zack, V. (2003). Navigating between theory and practice. Teachers who navigate between their research and their practice. Plenary panel. In N. A. Pateman, B. J. Dougherty & J. Zilliox (Eds.), *Proceedings PME 27* (Vol. 1, pp. 69–99). Hawai'i: Joseph: University of Hawai'i, CRDG, College of Education.

Novotná, J., Margolinas, C., & Sarrazy, B. (2013). Developing mathematics educators. In M. A. Clements, A. J. Bishop, C. Keitel, J. Kilpatrick & F. K. S. Leung (Eds.), *Third international handbook of mathematics education* (pp. 431–457). New York: Springer.

Novotná, J., & Moraová, H. (2005). Cultural and linguistic problems of the use of authentic textbooks when teaching mathematics in a foreign language. *ZDM Mathematics Education, 37*(2), 109–115.

Partová, E., & Marcinek, T. (2015). Third graders' representations of multiplication. In J. Novotná & H. Moraová (Eds.), *Proceedings of SEMT '15. Developing Mathematical Language and Reasoning* (pp. 267–275). Prague: Charles University, Faculty of Education.

Partová, E., Marcinek, T., Žilková, K., & Kopáčová, J. (2013). *Špecifické matematické poznatky pre vyučovanie [Specific mathematical knowledge for teaching]*. Bratislava: Vydavateľstvo UK.

Pavlovičová, G., & Barcíková, E. (2013). Investigation in geometrical thinking of pupils at the age of 11 to 12 through solving tasks. In J. Novotná & H. Moraová (Eds.), *Proceedings of SEMT '13. Tasks and Tools in Elementary Mathematics* (pp. 228–235). Prague: Charles University, Faculty of Education.

Polya, G. (1945). *How to solve it. A new aspect of mathematical method*. Princeton: Princeton University Press.

Rendl, M., Vondrová, N., et al. (2013). *Kritická místa matematiky na základní škole očima učitelů [Critical places of primary school mathematics in the eyes of teachers]*. Prague: Univerzita Karlova, Pedagogická fakulta.

Ripková, H., & Šedivý, J. (1986). Teorie vzdělávání v matematice [Theory of education in mathematics]. *Pokroky matematiky, fyziky, astronomie, 31*(6), 348–351.

Robová, J. (2012). *Informační a komunikační technologie jako prostředek aktivního přístupu žáků k matematice [Informational and communication technologies as a means of pupils' active approach to mathematics]*. Prague: Univerzita Karlova, Pedagogická fakulta.

Robová, J. (2013). Specific skills necessary to work with some ICT tools in mathematics effectively. *Didactica Mathematicae, 35,* 71–104.

Samková, L., & Hošpesová, A. (2015). Using concept cartoons to investigate future teachers' knowledge. In K. Krainer & N. Vondrová (Eds.), *Proceedings of CERME9* (3241–3247). Prague: Faculty of Education, Charles University and ERME.

Sedláček, L. (2009). A study of the influence of using dynamic geometry systems in mathematical education on the level of knowledge and skills of students. *Acta Didactica Universitatis Comenianae Mathematics, 9,* 81–108.

Semadeni, Z. (1985). *Nauczanie początkowe matematyki*. Warszawa: Wydawnictwa Szkolne i Pedagogiczne.

Shulman, L. S. (1986). Those who understand: knowledge growth in teaching. *Educational Researcher, 15*(2), 4–14. https://doi.org/10.3102/0013189x015002004.

Sikorová, Z. (2011). The role of textbooks in lower secondary schools in the Czech Republic. *IARTEM e-Journal, 4,* 1–22.

Simpson, A., & Stehlíková, N. (2006). Apprehending mathematical structure: A case study of coming to understand a commutative ring. *Educational Studies in Mathematics, 61*(3), 347–371. https://doi.org/10.1007/s10649-006-1300-y.

Simpson, A., Vondrová, N., & Žalská, J. (2018). Sources of shifts in pre-service teachers' patterns of attention: The roles of teaching experience and of observational experience. *Journal of Mathematics Teacher Education, 1*(6), 607–630. https://doi.org/10.1007/s10857-017-9370-6.

Slavíčková, M. (2007). Using educational software during mathematics lessons at lower secondary school. *Acta Didactica Universitatis Comenianae Mathematics, 7,* 117–129.

Smílková, J., & Balvín, J. (2016). *Jan Amos Komenský a jeho přínos filozofii výchovy a sociální pedagogice. [John Amos Comenius and his contribution to philosophy of education and social pedagogy.]*. Prague: Hnutí R.

Stehlíková, N. (2004). *Structural understanding in advanced mathematical thinking*. Prague: Charles University, Faculty of Education.

Steiner, H.-G., & Hejný, M. (Eds.) (1988). *Proceedings of the International Symposium on Research and Development in Mathematics Education*. Bratislava: Komenského Univerzita.

Tichá, M. (2000). Wie 11- bis 12-jährige Schüler Textaufgaben mit Brüchen begreifen. *Der Mathematikunterricht, 46*(2), 50–58.

Tichá, M., & Hošpesová, A. (2006). Qualified pedagogical reflection as a way to improve mathematics education. *Journal for Mathematics Teachers Education: Inter-Relating Theory and Practice in Mathematics Teacher Education, 9*(2), 129–156.

Tichá, M., & Hošpesová, A. (2013). Developing teachers' subject didactic competence through problem posing. *Educational Studies in Mathematics, 83*(1), 133–143.

Tichá, M., & Koman, M. (1998). On travelling together and sharing expenses examples of investigation of situations. *Teaching Mathematics and its Applications, 17*(3), 117–122.

Tůmová, V., & Vondrová, N. (2017). Links between success in non-measurement and calculation tasks in area and volume measurement and pupils' problems. *Scientia in educatione, 8*(2), 100–129.

Turnau, S. (1980). The mathematical textbook for young students. *Educational Studies in Mathematics, 11*(4), 393–410.

Vališová, A., & Kasíková, H. (Eds.). (2011). *Pedagogika pro učitele [Pedagogy for teachers]*. Prague: Grada.

Vaníček, J. (2009). *Počítačové kognitivní technologie ve výuce geometrie [Computer cognitive technologies in the teaching of geometry]*. Prague: Univerzita Karlova, Pedagogická fakulta.

Vankúš, P. (2008). Games based learning in teaching of mathematics at lower secondary school. *Acta Didactica Universitatis Comenianae Mathematics, 8,* 103–120.

Varga, T. (1976). *Mathematik 1: flussdiagramme, lochkarten, warscheinlichkeit*. Budapest: Akadémiai Kiadó.

Vondrová, N., Novotná, J., Havlíčková, R. (2018). The influence of situational information on pupils' achievement in additive word problems with several states and transformations. *ZDM, Advance Online Placement*, 1–15, https://doi.org/10.1007/s11858-018-0991-8

Vondrová, N., Novotná, J., & Tichá, M. (2015). Didaktika matematiky: historie, současnost a perspektivy s důrazem na empirické výzkumy [Didactics of mathematics: History, present days and perspective with a focus on empirical research]. In I. Stuchlíková & T. Janík (Eds.), *Oborové didaktiky: vývoj - stav - perspektivy* (pp. 93–122). Brno: Masarykova univerzita.

Vondrová, N., Rendl, M., Havlíčková, R., Hříbková, L., Páchová, A., & Žalská, J. (2015). *Kritická místa matematiky v řešeních žáků [Critical places of mathematics in pupils' solutions]*. Prague: Nakladatelství Karolinum.

Vondrová, N., & Žalská, J. (2015). Ability to notice mathematics specific phenomena: What exactly do student teachers attend to? *Orbis scholae, 9*(2), 77–101.

Vyšín, J. (1972). *Methoden zur Lösung mathematischer Aufgaben*. Leipzig: Teubner.

Vyšín, J. (1973). Vědeckovýzkumná práce v teorii vyučování matematice [Scientific work in the theory of teaching mathematics]. *Pokroky matematiky, fyziky a astronomie, 18*(1), 32–38.

Vyšín, J. (1976). Genetická metoda ve vyučování matematice [Genetic methods in the teaching of mathematics]. *Matematika a fyzika ve škole, 6,* 582–593.

Vyšín, J. (1980). Čechovy podněty pro vyučování matematice [Čech's inputs for the teaching of mathematics]. *Pokroky matematiky, fyziky, astronomie, 25*(6), 313–317.

Wittmann, E. C. (1981). *Grundfragen des Mathematikunterrichts*. Wiesbaden: Vieweg.

Wittmann, E. C. (1995). Mathematics education as a "design science". *Educational Studies in Mathematics, 29,* 355–374.

Žilková, K., Gunčaga, J., & Kopáčová, J. (2015). (Mis)Conceptions about geometric shapes in pre-service primary teachers. *Acta Didactica Napocensia, 8*(1), 27–35.